Fundamentals of Mathematical Proof

Charles A. Matthews

Southeastern Oklahoma State University

Library of Congress Control Number: 2018905479

Typeset in LaTeX by the author.

Department of Mathematics, Southeastern Oklahoma State University, Durant, OK, 74701. Email address cmatthews@se.edu.

ISBN-13: 978-1717176707

ISBN-10: 1717176704

10   9   8   7   6   5   4   3   2   1

Contents

Preface

This text evolved out of a set of notes from a one-semester course called *Foundations of Mathematics* at Southeastern Oklahoma State University. This course is required of all mathematics and mathematics education majors and minors. It is a prerequisite for abstract algebra, topology, real analysis, and almost all of the upper-level undergraduate math courses. In *Foundations of Mathematics*, students are expected to learn how to read and write proofs and to develop a sense of mathematical reasoning. In addition, they discover what much of undergraduate mathematics is really about.

A knowledge of calculus is not absolutely a prerequisite for an understanding of the material presented here. At Southeastern Oklahoma State, students who were enrolled in the *Foundations* class had either completed a course in calculus or were enrolled in a calculus class at the same time.

A complete understanding of the topics in this textbook can almost never occur without spending the time to complete solutions to the exercises which follow each section. Most people would agree that after just reading a book on how to serve a tennis ball, an inexperienced tennis player would not immediately become an excellent server. In the same way that a beginning tennis player must practice a serve in order to understand what he or she read, the reader of this text must practice by working through the exercises.

Acknowledgements

I am grateful to Professors Karl Frinkle, Christopher Moretti, and Patrick Reardon for the many useful suggestions you have made while using these notes to teach the *Foundations of Mathematics* class at Southeastern. My students in the *Foundations* classes have also contributed to the continual revisions of the notes until their present form, and I sincerely thank you for the improvements to the text that resulted from your input.

Chapter 1

Symbolic logic

Mathematics takes us into the region of absolute necessity, to which not only the actual world, but every possible world, must conform.

-Bertrand Russell, philosopher and logician (1872-1970)

1.1 Propositions, connectives and truth tables

Mathematics is about certainty. Mathematicians feel uncomfortable at best when making a claim of which they are not absolutely confident. This confidence is achieved only through a rigorous pursuit of the precise meaning of the claim and the reasons why the claim must be factual.

The benefits of developing the skills to analyze statements rigorously extend to almost every area of our lives. The ability to communicate effectively is of utmost importance in today's society. Many claims we might hear in our daily lives can be analyzed mathematically. Not too long ago a friend of mine asked me how I would interpret a written proposal made by a lawyer in a divorce settlement. I have changed the names (obviously) to protect the parties involved, but otherwise I quote, using the exact punctuation and capitalization:

> Mr. X will pay Ms. Y $100,000 in cash by September 1, 2001, and $160,000 in cash by November 1, 2001 OR Mr. X will pay Ms. Y

$240,000 payable at the rate of $4000 per month for 60 months at no interest, with sufficient security.

How should this proposal be interpreted? The problem is that there are several different ways to interpret the proposal. At first glance, it might seem that Mr. X has two options: pay $100,000 on September 1, plus $160,000 on November 1, for a total of $260,000; or pay a total of $240,000 over a period of 5 years. But this can't be the true intention of the proposal, for nobody would pay $260,000 now when they could have the option of paying $240,000 later!

Another interpretation would give Mr. X a different set of options: pay $100,000 on September 1, plus $160,000 on November 1, for a total of $260,000; or pay $100,000 on September 1, plus $240,000 over the next 5 years, for a total of $340,000. This interpretation makes some sense, since one would expect to pay more if the payments were spread out over a period of 5 years.

A third interpretation is possible for Mr. X: pay $100,000 on September 1, plus an additional $60,000 on November 1, for a total of $160,000 by November 1; or pay a total of $240,000 over a period of 5 years. This interpretation makes more sense than the first interpretation, since again Mr. X would pay more if he made payments over a 5-year period.

Unfortunately, there is no way to tell what was the original intention of the proposal. However, by analyzing this and similar proposals, one can become more proficient at understanding precise statements and communicating effectively.

In mathematics, claims are made in the form of propositions. A **proposition** or a **statement** is an unambiguous, declarative sentence that is either **true** or **false** (but not both true and false, if that is possible). Here *unambiguous* means that the sentence cannot be interpreted in more than one way. The following sentences are all propositions:

- Every differentiable function is continuous.
- $2 + 3 = 6$.
- For every real number x, if $x^2 > 9$ then $x > 3$.
- The number $\sqrt{2}$ is an irrational number.

Perhaps what a proposition *is* can best be understood by noticing what it *is not*. First, no question can be a proposition, since questions are not declarative. Similarly, commands like *"Read this sentence"* are not propositions because commands cannot really be true or false. Sentences which can sometimes be true and sometimes be false are considered ambiguous, and are not propositions. For example, the sentence *"There is no x such that $x^2 < 0$"* is ambiguous unless the context of what the variable x can be is specified. (Certainly there is no real number x such that $x^2 < 0$, but the complex number i does satisfy the inequality.) Also, the sentence *"$x^2 + 2x + 1 = 0$"* is not a proposition, for it is true when $x = -1$ and false when $x \neq -1$. Finally, some declarative sentences cannot be true or false. Such a sentence is called a **paradox**. The sentence *"This sentence is false"* is a paradox, for it cannot be true and it cannot be false.

> **Note:** Questions, commands, ambiguous sentences and paradoxes are not propositions.

The **truth value** of a proposition is the assignment of the value **true** or **false** to the proposition. In order to determine the truth value of a given proposition, we might first see if we can decompose the proposition into simpler propositions connected with words like *and*, *or*, and *if*. The truth values of the simple propositions will determine the truth value of the compound proposition.

For example, the compound proposition *"I will go to the store and buy some eggs"* can be decomposed into the two simple propositions *"I will go to the store"* and *"I will buy some eggs"*. The word *and* connects the two simple propositions with the meaning that *both* simple propositions must be true in order for the compound proposition to be true. This is called the **conjunction** of the two simple propositions. The conjunction of two propositions p and q is denoted with the symbol $\wedge$. In order for the compound proposition $p \wedge q$ to be true, both the simple proposition p and the simple proposition q must be true. The truth value of $p \wedge q$ is thus determined by the truth values of p and q as in the following table:

p	q	$p \wedge q$
T	T	T
T	F	F
F	T	F
F	F	F

The table above is called a **truth table**. A truth table for a compound proposition is a table containing every possible arrangement of truth values for the simple propositions which comprise the compound proposition, along with the resulting truth values of the compound proposition. Since there are 2 simple propositions in the compound proposition $p \wedge q$, and each simple proposition has 2 possible truth values (T and F), the table lists the $2 \cdot 2 = 4$ possible arrangements of truth values of the simple propositions.

The **disjunction** of the propositions p and q is denoted by $p \vee q$. The symbol $\vee$ represents the word "or," in the sense that for $p \vee q$ to be true, p or q or both p and q must be true. This is sometimes called the *inclusive or*, since the cases where the disjunction is true includes the case where both p and q are true. The truth table for the disjunction is as follows:

p	q	$p \vee q$
T	T	T
T	F	T
F	T	T
F	F	F

The **negation** of the proposition p is denoted by $\sim p$. The negation of p must have the opposite truth value from that of p. The truth table for $\sim p$ is:

p	$\sim p$
T	F
F	T

Example 1.1.1. Let p and q represent the following simple propositions.

p: The number π is a complex number.
q: The number i is an irrational number.

Then the propositions $p \vee \sim q$ and $\sim (q \wedge p)$ represent the following compound propositions:

$p \vee \sim q$:   The number π is complex or i is not an irrational number.

$\sim (q \wedge p)$: It is false that both i is irrational and π is complex.

$\square$

Example 1.1.2. Construct truth tables for $p \wedge (q \vee r)$ and $(p \wedge q) \vee r$. Do these compound propositions have the same meaning?

Solution. Since there are $2 \cdot 2 \cdot 2 = 8$ possible arrangements of the truth values of 3 simple propositions, each truth table should have 8 rows:

p	q	r	$q \vee r$	$p \wedge (q \vee r)$
T	T	T	T	T
T	T	F	T	T
T	F	T	T	T
T	F	F	F	F
F	T	T	T	F
F	T	F	T	F
F	F	T	T	F
F	F	F	F	F

p	q	r	$p \wedge q$	$(p \wedge q) \vee r$
T	T	T	T	T
T	T	F	T	T
T	F	T	F	T
T	F	F	F	F
F	T	T	F	T
F	T	F	F	F
F	F	T	F	T
F	F	F	F	F

Note that it is easy to compare these two tables because we have used the same arrangement of truth values of p, q and r in the first three columns of the tables. Any arrangement can be used, but if different tables contain different arrangements then the tables become more difficult to compare. In this text we will use the following arrangement: for the rightmost letter (r in this example), start with T and alternate truth values, as in T F T F ... . For the letter immediately left of that letter, start with T and change truth values after every 2 rows, as in T T F F ... . For the letter immediately left of that, start with T and change after every 4 rows. If there were a letter to the left of that one, we would start with T and change after every 8 rows.

Since the truth values in the last columns of our truth tables above do not always match, the two compound propositions do not have the same meaning. $\square$

Since the compound propositions $p \wedge (q \vee r)$ and $(p \wedge q) \vee r$ have different meanings, the parentheses are necessary to avoid ambiguity. However, the negation is given precedence over the conjunction and disjunction. This means that the proposition $\sim p \wedge q$ should be interpreted as $(\sim p) \wedge q$, not as $\sim (p \wedge q)$.

The symbols $\wedge$, $\vee$ and $\sim$ are called **connectives**. Another connective is the symbol $\rightarrow$. The compound proposition $p \rightarrow q$ can be read aloud as "if p then q." This is called a **conditional** proposition. The **hypothesis** or **antecedent** of $p \rightarrow q$ is the proposition p; the **conclusion** or **consequent** is q. The truth value of a conditional proposition is determined by the truth values of the hypothesis and the conclusion as in the following truth table:

p	q	$p \rightarrow q$
T	T	T
T	F	F
F	T	T
F	F	T

The truth values of $p \rightarrow q$ are a little less intuitive than those of the conjunction, disjunction or negation. In order to remember the truth table for the conditional $p \rightarrow q$, it might help to remember an example. Suppose your friend walks up to you and says, *"If I win the lottery then I will give you $10,000."* When would you consider this proposition to be a lie? If your friend actually wins the lottery and gives you $10,000, then certainly he or she has not lied. Also, if your friend does not win the lottery, then they really never had the opportunity to make their proposition into a lie. The only case in which your friend has lied is in the case when they won the lottery and did not give you $10,000. Consequently, the only case in which the conditional proposition $p \rightarrow q$ is false is the case in which p is true and q is false.

It is important to note that when $p \rightarrow q$ is true, it is not necessarily the case that the truth of p is the *cause* of the truth of q. For example, the conditional proposition, *"If $1 > 0$, then $1/2 = 0.5$"* is considered a true statement because both the hypothesis and the conclusion are true, yet the fact that $1 > 0$ did not cause $1/2$ to equal 0.5.

In English there are many different ways to express the conditional proposition $p \rightarrow q$; other than *if p then q*, it can be expressed as:

- *if p, q*
- *q, if p*

- *when p, then q*
- *when p, q*
- *q whenever p*
- *p only if q*
- *p is sufficient for q*
- *q is necessary for p*
- *p implies q*

Make sure not to interchange the hypothesis p and the conclusion q in any of these expressions, for the meaning of the resulting proposition would be different.

The last connective that we will introduce here is the symbol $\leftrightarrow$. The proposition $p \leftrightarrow q$ is called a **biconditional** proposition; it is read aloud as "p if and only if q." The meaning of $p \leftrightarrow q$ is the same as that of $(p \rightarrow q) \wedge (q \rightarrow p)$. We can construct the truth table for $(p \rightarrow q) \wedge (q \rightarrow p)$ in order to determine the truth values for $p \leftrightarrow q$:

p	q	$p \rightarrow q$	$q \rightarrow p$	$(p \rightarrow q) \wedge (q \rightarrow p)$
T	T	T	T	T
T	F	F	T	F
F	T	T	F	F
F	F	T	T	T

Hence, the truth table for $p \leftrightarrow q$ is

p	q	$p \leftrightarrow q$
T	T	T
T	F	F
F	T	F
F	F	T

An easy way to remember the truth values of $p \leftrightarrow q$ is to interpret the proposition $p \leftrightarrow q$ to mean, "p and q have the same truth value." Note how the truth values in the above table fit this interpretation; $p \leftrightarrow q$ is true when p and q are both true or both false, and $p \leftrightarrow q$ is false when p and q have different truth values.

When different connectives are used in one compound proposition, a standard order of precedence should be observed. Unless parentheses are used to force connectives to be applied in a particular order, the negation is always applied first, followed by the conjunction and the disjunction, then the conditional, and finally the biconditional. The following table summarizes the connectives introduced so far.

connective	name	usage	spoken form	precedence
$\sim$	negation	$\sim p$	not p	first
$\wedge$	conjunction	$p \wedge q$	p and q	second
$\vee$	disjunction	$p \vee q$	p or q	
$\rightarrow$	conditional	$p \rightarrow q$	if p then q	third
$\leftrightarrow$	biconditional	$p \leftrightarrow q$	p if and only if q	last

In general, a truth table for a compound proposition which contains n distinct simple propositions and m connectives will have 2^n rows and $n + m$ columns. The first n columns should have the simple propositions for their headings, and there should be a new column for each connective. Each column heading should be a proposition like "$p \wedge q$," rather than just "$\wedge$," which is not a proposition by itself. The last column heading should be the compound proposition for which we are constructing the truth table.

For example, the truth table for the compound proposition

$$\sim (p \vee q) \rightarrow (\sim p \leftrightarrow r)$$

should contain $2^3 = 8$ rows and $3 + 5 = 8$ columns. The column headings should appear as:

$$p \mid q \mid r \mid p \vee q \mid \sim (p \vee q) \mid \sim p \mid \sim p \leftrightarrow r \mid \sim (p \vee q) \rightarrow (\sim p \leftrightarrow r)$$

The truth values in the table should be filled in one column at a time, working from left to right.

For more complicated propositions, it sometimes helps to insert as many sets of parentheses into the proposition as possible, using the rules of precedence for the connectives. For example, consider the proposition

$$\sim p \vee q \rightarrow r \leftrightarrow p \wedge \sim q \rightarrow r.$$

Since the negation has first precedence, we first insert parentheses for the negations:

$$(\sim p) \vee q \rightarrow r \leftrightarrow p \wedge (\sim q) \rightarrow r.$$

Next we insert parentheses for the conjunction and disjunction:

$$((\sim p) \vee q) \rightarrow r \leftrightarrow (p \wedge (\sim q)) \rightarrow r.$$

Then we insert parentheses for the conditionals:

$$(((\sim p) \vee q) \to r) \leftrightarrow ((p \wedge (\sim q)) \to r).$$

Hence, we can use the following column headings in the truth table for this proposition, where for the sake of space we use the symbol * to represent the whole proposition:

p	q	r	$\sim p$	$\sim p \vee q$	$\sim p \vee q \to r$	$\sim q$	$p \wedge \sim q$	$p \wedge \sim q \to r$	*

Alternatively, we can use the following column headings for the same proposition:

p	q	r	$\sim p$	$\sim q$	$\sim p \vee q$	$p \wedge \sim q$	$\sim p \vee q \to r$	$p \wedge \sim q \to r$	*

The first arrangement of column headings is slightly preferable to the second because it progresses through the proposition from left to right as much as possible, whereas the second arrangement skips back and forth through the proposition a bit too much.

If a column heading is a negation, then the truth values in that column depend on only one of the previous columns; otherwise, the truth values in that column depend on two of the previous columns. The basic truth tables for the five connectives studied in this section should be memorized in order to be able to fill in larger truth tables in a timely manner.

Exercises

1.1.1. The following quote appeared on a course syllabus for Math 3133.

> **PREREQUISITE**: The student must have completed the following or the equivalent:
> (1) Math 1303 – Mathematics in the Liberal Arts; and
> (2) Math 1513 – College Algebra; OR
> (3) Math 1543 – Algebra for the Sciences.

List two different interpretations of this policy. For each interpretation, clearly state the options for the prerequisites of Math 3133.

1.1.2. Which of the following are propositions? Justify your answer.
(a) Read this sentence.

(b) There is a real number x such that $3x^5 - 4x^4 + 5x^3 - 6x^2 + 7x + 9 = 0$.

(c) I am lying.

(d) Every real number is a complex number.

(e) The number 2 is odd.

(f) The derivative of e^{x^2} is e^{x^2}.

(g) Can you take the derivative of any polynomial?

(h) Find the derivative of $\cos x^2$.

(i) $x > 1$.

1.1.3. Explain thoroughly why the sentence "*This sentence is false*" is a paradox.

1.1.4. Construct truth tables for the propositions $p \vee (q \vee r)$ and $(p \vee q) \vee r$. Do these compound propositions have the same meaning?

1.1.5. Construct truth tables for the propositions $p \wedge (q \wedge r)$ and $(p \wedge q) \wedge r$. Do these compound propositions have the same meaning?

1.1.6. Construct truth tables for the propositions $p \to (q \to r)$ and $(p \to q) \to r$. Do these compound propositions have the same meaning?

1.1.7. Construct truth tables for the propositions $p \leftrightarrow (q \leftrightarrow r)$ and $(p \leftrightarrow q) \leftrightarrow r$. Do these compound propositions have the same meaning?

1.1.8. How many rows and columns should be included in a truth table for each of the following propositions? What should the column headings be?

(a) $p \wedge \sim p$

(b) $p \wedge \sim q$

(c) $p \wedge (p \vee \sim q) \wedge (r \vee \sim s)$

(d) $\sim p \leftrightarrow p \wedge q \to (\sim r \vee p)$

1.1.9. The disjunction $\vee$ is occasionally called the "inclusive or" because $p \vee q$ is considered true when p or q is true, including the case when both p and q are true. Let $\underline{\vee}$ denote the "exclusive or," so that $p\underline{\vee}q$ is true if and only if exactly one of them is true. Construct the truth table for $p\underline{\vee}q$.

1.1.10. Construct a truth table for the proposition

$$p \vee q \to r \leftrightarrow (p \to r) \wedge (q \to r).$$

1.1.11. Construct a truth table for the proposition

$$p \vee q \leftrightarrow r \to \sim s \wedge q.$$

1.1.12. If the hypothesis and the conclusion of a conditional proposition are interchanged, does the resulting proposition have the same meaning as the original conditional proposition? Justify your answer with truth tables.

1.1.13. Determine whether each of the following propositions is true or false.
(a) $2 + 4 = 5$ or $6^2 < 70$.
(b) $7 - 3 = 4$ or $5 + 4 = 9$.
(c) If $1 + 3 = 5$ then $7 - 2 = 5$.
(d) If $13 + 7 = 20$ then $13 - 7 = 6$.
(e) If $2 \cdot 4 = 6$ then $7 + 3 = 9$.
(f) If $2 - 3 = -1$ then $3 - 3 = 2$.
(g) If $3 + 2 = 4$ or $2 + 5 = 7$, then $3 + 3 = 5$ and $4 + 4 = 8$.
(h) $3 + 3 = 6$ if and only if $4 + 4 = 8$.
(i) $3 - 2 = 2$ if and only if $70 - 30 = 40$.
(j) $1 + 2 = 4$ if and only if $1 + 2 = 5$.

1.1.14. A somewhat famous joke about mathematicians goes as follows:

> While on a train traveling in Scotland for the first time, an astronomer, a physicist and a mathematician see a black sheep in a pasture. The astronomer says, "Look! Scottish sheep are black!" The physicist answers, "Well, one Scottish sheep is black." The mathematician replies, "Well, one Scottish sheep is black on one side."

What is it about the nature of mathematicians that this joke is pointing out? (Whether the joke is actually funny is another matter.)

1.2 Logical equivalence and implication

The truth values of most compound propositions depend on the truth values of the simple propositions that are contained in the compound propositions. But it is possible that a compound proposition is always true or always false, regardless of the truth values of the simple propositions involved. For example, regardless of the truth value of the simple proposition p, the compound proposition $p \wedge \sim p$ is always false. A **contradiction** is a compound proposition which is always false. A short truth table verifies that $p \wedge \sim p$ is a contradiction:

p	$\sim p$	$p \wedge \sim p$
T	F	F
F	T	F

Note that the last column consists of only F's. Another contradiction is the proposition $\sim p \wedge \sim (p \to q)$.

A **tautology** is a compound proposition which is always true. For example, the compound proposition $p \vee \sim p$ is a tautology. Think about why this must be a tautology; the claim, "*I will make an A, or I will not make an A*" must always be true.

Let P and Q be (possibly compound) propositions. (Note that we are using capital letters instead of the lower-case p and q here. We will use capitals for our propositions when they can be compound propositions, as opposed to only simple ones.) Propositions P and Q are **logically equivalent** if the biconditional proposition $P \leftrightarrow Q$ is a tautology. What this means is that the truth values of P and Q are the same, so that P and Q have the same meaning. When P and Q are logically equivalent, we write $P \Leftrightarrow Q$. (Some textbooks write $P \equiv Q$ here, but we will reserve the symbol $\equiv$ for modular congruence, which will be defined later.) The difference between $P \leftrightarrow Q$ and $P \Leftrightarrow Q$ is that $P \leftrightarrow Q$ is a proposition which can be either true or false; on the other hand, we write $P \Leftrightarrow Q$ to indicate that the biconditional proposition $P \leftrightarrow Q$ is a tautology.

Example 1.2.1. Show that $(p \to q) \Leftrightarrow (\sim q \to \sim p)$.

Solution. We construct a truth table to show that $(p \to q) \leftrightarrow (\sim q \to \sim p)$ is a tautology:

p	q	$p \to q$	$\sim q$	$\sim p$	$\sim q \to \sim p$	$(p \to q) \leftrightarrow (\sim q \to \sim p)$
T	T	T	F	F	T	T
T	F	F	T	F	F	T
F	T	T	F	T	T	T
F	F	T	T	T	T	T

$\square$

The conditional proposition $\sim q \to \sim p$ is called the **contrapositive** of the conditional proposition $p \to q$. Example 1.2.1 shows that a conditional proposition and its contrapositive are logically equivalent. Thus, the statements "*If Joe knows math, then Joe knows calculus*" and "*If Joe doesn't know calculus, then Joe doesn't know math*" have the same meaning.

The **converse** of $p \rightarrow q$ is the proposition $q \rightarrow p$. A conditional statement and its converse are not logically equivalent. The converse of *"If Joe has a Ferrari, then Joe has a car"* is *"If Joe has a car, then Joe has a Ferrari."* Most people would agree that the first proposition is true, whereas the second is false; so the conditional proposition can have a different truth value than its converse.

The **inverse** of $p \rightarrow q$ is $\sim p \rightarrow \sim q$. The inverse of a conditional proposition is the contrapositive of the converse of the proposition. Therefore, the inverse and the converse are logically equivalent; but neither is logically equivalent to the original conditional proposition. On the other hand, neither is logically equivalent to the negation of $p \rightarrow q$; that is, neither $q \rightarrow p$ nor $\sim p \rightarrow \sim q$ is logically equivalent to $\sim (p \rightarrow q)$.

Example 1.2.2. Show that $\sim (p \wedge q)$ and $\sim p \wedge \sim q$ are not logically equivalent.

Solution. We construct a truth table to show that $\sim (p \wedge q) \leftrightarrow \sim p \wedge \sim q$ is not a tautology:

p	q	$p \wedge q$	$\sim (p \wedge q)$	$\sim p$	$\sim q$	$\sim p \wedge \sim q$	$\sim (p \wedge q) \leftrightarrow \sim p \wedge \sim q$
T	T	T	F	F	F	F	T
T	F	F	T	F	T	F	F
F	T	F	T	T	F	F	F
F	F	F	T	T	T	T	T

$\square$

It is left to the exercises to show that $\sim (p \vee q)$ and $\sim p \vee \sim q$ are not logically equivalent. According to the following theorem, the negation of a conjunction is the disjunction of the negations, and the negation of a disjunction is the conjunction of the negations.

Theorem 1.2.3 (DeMorgan's Laws for Logic). (a) $\sim (p \wedge q) \Leftrightarrow \sim p \vee \sim q$. (b) $\sim (p \vee q) \Leftrightarrow \sim p \wedge \sim q$.

The proof of DeMorgan's Laws is left to the exercises. To give an example of DeMorgan's Laws in English, the meaning of *"It is not true that I will get an A in biology and an A in physics"* is the same as that of *"I will not get an A in biology or I will not get an A in physics."*

Proposition 1.2.4. *The propositions $p \vee q$ and $\sim p \rightarrow q$ are logically equivalent.*

Proof. The truth table below verifies that $(p \vee q) \leftrightarrow (\sim p \to q)$ is a tautology.

p	q	$p \vee q$	$\sim p$	$\sim p \to q$	$(p \vee q) \leftrightarrow (\sim p \to q)$
T	T	T	F	T	T
T	F	T	F	T	T
F	T	T	T	T	T
F	F	F	T	F	T

$\square$

Proposition 1.2.5. *The conditional proposition $p \to q$ is logically equivalent to the disjunction $\sim p \vee q$.*

The proof of Proposition 1.2.5 is Exercise 1.2.8.

Proposition 1.2.6. *The proposition $\sim (p \to q)$ is logically equivalent to $p \wedge \sim q$.*

Proof. A truth table can be used to show that $\sim (p \to q) \leftrightarrow (p \wedge \sim q)$ is a tautology, but there is an alternate verification that we can use here. We can apply Proposition 1.2.5 to conclude that $\sim (p \to q)$ is logically equivalent to $\sim (\sim p \vee q)$. Then we can apply DeMorgan's Law (Theorem 1.2.3 (b))to get $\sim (\sim p \vee q) \Leftrightarrow (\sim (\sim p)) \wedge (\sim q)$. Since $\sim (\sim p)$ is clearly logically equivalent to p, we can conclude that $\sim (p \to q) \Leftrightarrow p \wedge \sim q$ by following the string of logical equivalencies

$$\sim (p \to q) \Leftrightarrow \sim (\sim p \vee q) \Leftrightarrow (\sim (\sim p) \wedge (\sim q)) \Leftrightarrow (p \wedge \sim q).$$

$\square$

An **implication** is a conditional proposition $P \to Q$ which is a tautology. If $P \to Q$ is an implication, we write $P \Rightarrow Q$. For example, since the conditional proposition $(p \wedge q) \to p$ is a tautology, it is an implication, and we write $(p \wedge q) \Rightarrow p$. Similarly, $p \Rightarrow (p \vee q)$. On the other hand, $(p \vee q) \to p$ is a conditional statement, and since it is not a tautology, it is not an implication. It would be incorrect to write "$(p \vee q) \Rightarrow p$," but it is okay to write "$(p \vee q) \to p$," since this is just a proposition and there is no claim that this proposition is always true.

There are two basic types of reasoning: inductive and deductive reasoning. **Inductive reasoning** is the kind of reasoning in which conclusions are based on experience, experimentation, and past events. For example, we might conclude that the sun will rise tomorrow based on the evidence that the sun has risen every day that we can remember. We might conclude that when we mix certain chemicals, an explosion will always result, based upon the evidence that when these chemicals were mixed in the past, explosions always occurred.

When inductive reasoning is used, we can never be completely certain that the conclusions are valid. Indeed, some event might occur that would prevent Earth from rotating, and the sun might not rise tomorrow. Likewise, when chemicals were mixed in the past and explosions always occurred, it is possible that other factors were involved that led to the explosions. If the same chemicals are mixed at a low temperature, perhaps an explosion will not result.

Example 1.2.7. Make a chart of values of $n^2 - n + 11$ for some positive integers n, and use inductive reasoning to make a conclusion concerning these values.

Solution. Let's make a chart for integers n from 1 to 9:

n	$n^2 - n + 11$
1	11
2	13
3	17
4	23
5	31
6	41
7	53
8	67
9	83

We might make several conclusions about the values of $n^2 - n + 11$. First, we might say that the value of $n^2 - n + 11$ is always odd. We might also say that the last digit is always 1, 3, or 7. Furthermore, we might notice that all the values of $n^2 - n + 11$ in the chart are prime, so we might conclude that $n^2 - n + 11$ is always prime. (An integer m is **prime** if $m > 1$ and the only positive integers that divide m are 1 and m. An integer a **divides** the integer b if there is an integer c such that $ac = b$. For example, 6 is not prime since 2 divides 6. We will talk more about this soon.)

But just because $n^2 - n + 11$ is odd, ends in 1, 3, or 7, and is prime for $1 \leq n \leq 9$, does this mean that this will always be the case? We might try to see if these conclusions work for $n = 10$. When $n = 10$, $n^2 - n + 11 = 101$, which again is odd, ends in 1, and is prime. How many more values of n would it take before you would be convinced? 10? 100? 1,000,000?

Actually, if we try just one more value of n, we find that one of our conclusions was not correct. When $n = 11$, $n^2 - n + 11 = 121$, and 121 is not prime, since 11 divides 121.

Hopefully, you can see the problem with inductive logic here: no matter how many values of n you check, you can never be absolutely certain that your conclusions will always be valid. □

Deductive reasoning is the kind of reasoning in which conclusions are based on accepted assumptions. These accepted assumptions are sometimes called **premises**. In deductive reasoning, if an argument is logically valid and the premises are accepted, then the conclusion must be accepted also.

Often, inductive reasoning gets us to the conclusion, and deductive reasoning is used to prove the conclusion with certainty.

Symbolic deductive arguments follow a standard pattern, where the premises are listed first, followed by a line and the conclusion. For example, in the symbolic deductive argument

$$p \to q$$
$$\underline{p \qquad\qquad}$$
$$\therefore q$$

there are two premises: $p \to q$ and p. These are the assumptions of the argument, meaning that we are assuming that these two propositions are actually true. The symbol $\therefore$ represents the word *therefore*, and the conclusion is that the proposition q must be true.

Every deductive argument has at least one premise and a conclusion. We can write the argument as a long conditional proposition which connects the premises with conjunctions and has the conclusion of the argument as the conclusion of the conditional proposition. The argument above can be written as

$$(p \to q) \wedge p \to q.$$

Notice that we use the conjunction $\wedge$ in between the premises, since we are assuming that $p \to q$ AND p are true.

A deductive argument is **logically valid** if it is an implication. To prove that the argument

$$(p \to q) \wedge p \to q$$

is logically valid, we can construct a truth table to show that this conditional proposition is a tautology and therefore

$$(p \to q) \wedge p \Rightarrow q.$$

This deductive argument is well-known and has the name **modus ponens** from the Latin phrase meaning *mode that affirms*. An example of modus ponens is the English argument

> *If Joe studies hard, then Joe will get an A on the test.*
> *Joe will study hard.*
> *Therefore, Joe will get an A on the test.*

This argument is logically valid. But it is not necessarily true that Joe will get an A on the test! What is true is that *if* both premises are true, *then* Joe will get an A on the test.

A deductive argument is called **sound** if it is valid and all of the premises are true. One should not confuse the validity and the soundness of an argument.

Example 1.2.8. Determine whether the following argument is logically valid.

> *If Alaina wins the race, then Alaina will be happy.*
> *Alaina will not win the race.*
> *Therefore, Alaina will not be happy.*

Solution. The first step is to choose letters to represent the simple propositions involved in the argument. Let w represent the proposition, "*Alaina will win the race.*" Let h represent, "*Alaina will be happy.*" Then the symbolic form of the deductive argument is

$$\begin{array}{c} w \to h \\ \sim w \\ \hline \therefore \sim h \end{array}$$

As a conditional proposition, the argument can be written as

$$(w \to h) \wedge \sim w \to \sim h.$$

In order to determine whether the argument is logically valid, we construct a truth table:

w	h	$w \to h$	$\sim w$	$(w \to h) \wedge \sim w$	$\sim h$	$(w \to h) \wedge \sim w \to \sim h$
T	T	T	F	F	F	T
T	F	F	F	F	T	T
F	T	T	T	T	F	F
F	F	T	T	T	T	T

Since the conditional statement is not a tautology, it is not an implication, and therefore the argument is not valid. In fact, the truth table points out the case in which the argument fails: if w is false and h is true, then the conditional statement is false. That is, in the case where Alaina does not win the race, but she is happy anyway, both premises are true, yet the conclusion is not. □

Example 1.2.9. The deductive argument

$$p \to q$$
$$\frac{\sim q}{\therefore \sim p}$$

is logically valid. It is called **modus tollens** (Latin for *mode that denies*). An example of modus tollens in English is the following argument:

> *If I have a Ferrari, then I have a car. I don't have a car. Therefore, I don't have a Ferrari.*

□

Example 1.2.10. Another commonly used valid argument is

$$p \to q$$
$$\frac{q \to r}{\therefore p \to r}$$

This argument is called the **hypothetical syllogism**. One can add more premises to the hypothetical syllogism to get similar valid arguments like

$$p \to q$$
$$q \to r$$
$$\frac{r \to s}{\therefore p \to s}$$

□

Example 1.2.11. A truth table can be used to verify that the argument

$$p \to q$$
$$\underline{\quad q \quad}$$
$$\therefore p$$

is not valid. An argument which is not logically valid is called a **fallacy**. Thinking of this argument as the conditional proposition $(p \to q) \land q \to p$, we can see that when p is false and q is true, the hypothesis is true and the conclusion is false. This is the only time when a conditional proposition is false (when the hypothesis is true but the conclusion is false). In general, if it is possible that all the premises of an argument are true and the conclusion is false, then the argument is a fallacy.

The argument

$$\underline{p \to q}$$
$$\therefore q$$

is also a fallacy, since it is possible that $p \to q$ is true and q is false. (This happens in the case when both p and q are false.) $\square$

Exercises

1.2.1. Verify that $\sim p \land \sim (p \to q)$ is a contradiction.

1.2.2. Verify that $\sim q \lor (p \to q)$ is a tautology.

1.2.3. (a) Show that $q \to p$ is not logically equivalent to $\sim (p \to q)$.
(b) Show that $\sim p \to \sim q$ is not logically equivalent to $\sim (p \to q)$.

1.2.4. Write English sentences which form the contrapositive, the converse, and the inverse of the conditional proposition, "*If $f(x) = x^2$, then $f'(x) = 2x$.*"

1.2.5. Verify that $\sim (p \lor q)$ and $\sim p \lor \sim q$ are not logically equivalent.

1.2.6. Use truth tables to prove DeMorgan's Laws (Theorem 1.2.3).

1.2.7. Express in English the contrapositive, the converse, and the inverse of the conditional proposition, "*If $f'(x) = 2x$ and $f(0) = 4$, then $f(1) = 5$ or $f'(1) = 5$.*" Use DeMorgan's Laws whenever possible.

1.2.8. Use a truth table to verify Proposition 1.2.5.

1.2.9. Construct a truth table or truth tables to determine whether the propositions $(p \wedge q) \to r$ and $(p \to r) \wedge (q \to r)$ are logically equivalent.

1.2.10. (a) Use truth tables to verify that $p \to (q \vee r) \Leftrightarrow (p \wedge \sim q) \to r$. (b) Use Proposition 1.2.5 to give an alternate verification (using a string of logical equivalencies as in the proof of Proposition 1.2.6) that $p \to (q \vee r) \Leftrightarrow (p \wedge \sim q) \to r$.

1.2.11. How many non-equivalent compound propositions are there that use only two simple propositions p, q? How many that use three? How many that use n simple propositions?

1.2.12. Verify that $(p \wedge q) \Rightarrow p$ and $p \Rightarrow (p \vee q)$.

1.2.13. Use inductive reasoning to make some conclusions about the values of $n^2 - n + 13$.

1.2.14. Goldbach's Conjecture states that every even integer greater than 4 can be written as the sum of two odd prime integers. Christian Goldbach first made this conjecture in a letter to Leonard Euler in 1742. The conjecture was verified for all integers up to 10^{14} in the year 1998, and the verification continues today. A \$1 million prize was offered recently for any correct proof of Goldbach's Conjecture. Verify the conjecture for all integers up to 30.

1.2.15. Verify that the modus tollens deductive argument (given in Example 1.2.9) is logically valid.

1.2.16. Verify that the hypothetical syllogism (given in Example 1.2.10) is logically valid.

1.2.17. Verify that the argument

$$\begin{array}{c} p \vee q \\ \underline{\sim p} \\ \therefore q \end{array}$$

is logically valid.

1.2.18. Use a truth table to verify that the argument

$$\begin{array}{c} p \to q \\ \underline{q} \\ \therefore p \end{array}$$

is not valid. Point out the case in which the argument fails.

1.2.19. Use a truth table to verify that the argument

$$\frac{p \rightarrow q}{\therefore q}$$

is not valid. Point out the case in which the argument fails.

1.2.20. What conclusions can be logically deduced based on the following list of premises?

$$\begin{array}{c} p \rightarrow q \\ q \rightarrow r \\ \sim s \rightarrow \sim r \\ s \rightarrow t \\ \dfrac{\sim t}{\therefore \ ?} \end{array}$$

1.2.21. Determine whether the following argument is logically valid.

If my teacher is interesting, then I stay awake. I stay awake in class. Therefore, my teacher is interesting.

1.2.22. Determine whether the following argument is logically valid.

If I open the box, then it will explode. I will not open the box. Therefore, it will not explode.

1.2.23. Determine whether the following argument is logically valid.

If I go to the store, then I will buy some ice cream. We will have a party if I buy some ice cream. If we have a party then we will watch the game. Therefore, if we watch the game, then I will go to the store.

1.2.24. Given the following assumptions, determine who shot the sheriff. Then prove your answer by writing a deductive argument in symbolic form and checking that the argument is logically valid.

If Alex shot the sheriff, then Alex shot Roman.
If Alex did not shoot the sheriff, then Roman shot the sheriff.
If Roman shot the sheriff, then the sheriff did not shoot the deputy.
The sheriff shot the deputy.

1.2.25. Consider the following deductive argument.

> *If* $2 > 3$, *then* $3 > 4$.
> *If* $3 > 4$, *then* $9 > 16$.
> $2 > 3$.
> *Therefore,* $9 > 16$.

(a) Is this argument valid?
(b) Is this argument sound?

1.3 Quantifiers

Previously, we asserted that a proposition must be unambiguous, and that a statement like "$4x^2 - 4x + 1 = 0$" is not a proposition because it is true when $x = 1/2$ and false when $x = 2$. A statement containing a variable that becomes a proposition when the variable takes on values is called a **predicate** or a **propositional function**. For example, we can let $p(x)$ denote the predicate "$4x^2 - 4x + 1 = 0$", and then $p(1/2)$ is a true proposition and $p(2)$ is a false proposition.

One way to make a proposition from a predicate is to substitute values for the variable, but this is not the only way. Another way is to *quantify* the predicate. For example, the predicate $4x^2 - 4x + 1 = 0$ can be quantified by stating,

> *There exists an x such that $4x^2 - 4x + 1 = 0$.*

Mathematicians use a backwards **E** to denote the quantification "*There exists a*" or "*There is some.*" Hence, the above proposition can be written symbolically as

$$\exists x \; (4x^2 - 4x + 1 = 0).$$

The symbol $\exists$ is called the **existential quantifier**.

Since the value $x = 1/2$ satisfies the equation $4x^2 - 4x + 1 = 0$, the truth value of the proposition $\exists x \; (4x^2 - 4x + 1 = 0)$ is true. On the other hand, consider the proposition $\exists x \; (x^2 + x + 1 = 0)$. Since the equation $x^2 + x + 1 = 0$ has no real solutions, one might consider the proposition to be false. But the equation $x^2 + x + 1 = 0$ does have two complex solutions:

$$x = \frac{-1 \pm \sqrt{-3}}{2}.$$

Does this make the proposition $\exists x \ (x^2 + x + 1 = 0)$ true? The answer to this question depends on whether we are looking only for real solutions or we are willing to accept complex solutions. What kind of values we are willing to accept for the variable should be made clear before a quantified predicate can be assigned a truth value.

The **domain of discourse** (or the **universe of discourse**) of a predicate is the set of all values for the variable in the predicate that we wish to consider. By changing the domain of discourse, we might change the truth value of a quantified predicate.

There is a standard notation for many of the common domains of discourse for numerical variables:

$\mathbb{C}$	the set of all complex numbers
$\mathbb{R}$	the set of all real numbers
$\mathbb{R}^+$	the set of positive real numbers
$\mathbb{Q}$	the set of all rational numbers
$\mathbb{Q}^+$	the set of positive rational numbers
$\mathbb{Z}$	the set of all integers
$\mathbb{N}$	the set of all natural numbers

A **complex number** is a number which can be expressed in the form $a + bi$ where a and b are real numbers and i denotes a square root of -1. A **rational number** is a real number that can be expressed as a quotient of integers n/m, where $m \neq 0$. (If a number is real but not rational, it is called **irrational**.) A real number is called **positive** if it is greater than 0; it is called **negative** if it is less than 0. (Hence, the number 0 is neither positive nor negative.) A **natural number** is a positive integer.

The symbol $\in$ can be used to indicate inclusion in one of these domains of discourse. Thus, we write

$$x \in \mathbb{R}$$

to indicate that x must be a real number, or

$$x \in \mathbb{N}$$

to indicate that x must be a natural number. We can use this notation to specify a particular domain of discourse. Hence, the proposition

$$\exists x \in \mathbb{R} \ (x^2 + x + 1 = 0)$$

is false but the proposition

$$\exists x \in \mathbb{C} \ (x^2 + x + 1 = 0)$$

is true. Note that the last proposition can be read aloud as

There is a complex number x such that $x^2 + x + 1 = 0$.

When the domain of discourse is clear from the context, it can be left out of the symbolic expression for the quantified predicate. For example, if it is clear that the domain of discourse is the set of real numbers, then the meaning of

$$\exists x \ (x^2 = 5)$$

is evident.

Example 1.3.1. Let $p(x)$ denote the predicate "$5x = 2$." Then the proposition

$$\exists x \in \mathbb{N} \ p(x)$$

is false, but

$$\exists x \in \mathbb{Q} \ p(x)$$

is true.   □

When the domain of discourse for a predicate $p(x)$ consists of finitely many values, the quantified predicate $\exists x \ p(x)$ is logically equivalent to a proposition involving disjunctions. For example, when the domain of discourse for $p(x)$ consists of only the numbers 0, 1 and 2, then $\exists x \ p(x)$ is logically equivalent to $p(0) \vee p(1) \vee p(2)$. When the domain of discourse consists of infinitely many values, then infinitely many disjunctions would be needed to rephrase the existentially quantified predicate.

The **exclamation point** is used after the symbol $\exists$ to denote uniqueness; so the proposition $\exists! x \, p(x)$ means *There is one and only one x such that $p(x)$*, or *There is a unique x such that $p(x)$*. For example,

$$\exists x \in \mathbb{R} \ (x + 2)(x - 3) = 0$$

is a true proposition because there is a real number x which makes the product $(x + 2)(x - 3)$ equal to 0; but

$$\exists! \, x \in \mathbb{R} \ (x + 2)(x - 3) = 0$$

is a false proposition because there is more than one real number x which makes the product 0. However, the proposition

$$\exists! \, x \in \mathbb{N} \, (x+2)(x-3) = 0$$

is true since the only natural number satisfying the equation is $x = 3$.

In symbolic terms, $\exists! x \, p(x)$ can be defined as:

$$(\exists x \, p(x)) \wedge (\sim \exists x_1, x_2 \, (x_1 \neq x_2) \wedge p(x_1) \wedge p(x_2) \,).$$

In English, this expression means that there exists a value x which makes $p(x)$ true, and there do not exist two distinct values x_1, x_2 such that both $p(x_1)$ and $p(x_2)$ are true.

Another way to quantify a predicate is by proposing that the predicate is true for *every* value in the domain of discourse:

$$\text{For all } x \, p(x).$$

The up-side-down letter A is used to denote this quantification:

$$\forall x \ p(x).$$

The symbol $\forall$ is called the **universal quantifier**. In order for the proposition $\forall x \ p(x)$ to be true, $p(x)$ must be true for every value of x in the domain of discourse.

Example 1.3.2. Let $p(x)$ denote the predicate "$x^2 \geq 0$" and let $q(x)$ denote the predicate "$x^2 > 0$. Then

$$\forall x \in \mathbb{R} \ p(x)$$

is a true proposition, but

$$\forall x \in \mathbb{R} \ q(x)$$

is false because 0 is a real number and $q(0)$ is false. $\square$

Remark 1.3.3. Most predicates involving conditional or biconditional statements are assumed to be universally quantified. For example, the predicate "*If $x > 5$ then $x^2 > 25$*" is used with the implied meaning of the universally quantified proposition, "*For all real numbers x, if $x > 5$ then $x^2 > 25$.*" Hopefully the domain of discourse will be clear from the context. $\square$

Just as an existentially quantified predicate can be logically equivalent to a proposition involving disjunctions, a universally quantified predicate can be logically equivalent to a proposition involving conjunctions. For example, when the domain of discourse consists of only the numbers 0, 1 and 2, then $\forall x\ p(x)$ is logically equivalent to $p(0) \wedge p(1) \wedge p(2)$.

Example 1.3.4. Write each of the following in symbolic form.
(a) Every differentiable function is continuous.
(b) Some continuous functions are differentiable.
(c) No real number has a square of -9.

Solution. (a) This proposition is universally quantified, but it might not be obvious at first how to construct the predicate. The meaning of the proposition is that if a function is differentiable, then it is continuous. So we see that the predicate here is a conditional predicate. Let $d(f)$ denote the predicate, "*The function f is differentiable,*" and let $c(f)$ denote the predicate, "*The function f is continuous.*" Then the symbolic form of the quantified conditional predicate is

$$\forall f\ d(f) \to c(f).$$

The domain of discourse here is the set of all (real-valued) functions.

(b) This proposition is existentially quantified, and its meaning is the same as that of, "*There is some function f such that f is continuous and f is differentiable.*" Letting $d(f)$ and $c(f)$ denote the same predicates we used in part (a), this proposition has the symbolic form

$$\exists f\ c(f) \wedge d(f).$$

(c) We can rephrase this proposition as, "*It is false that there is a real number x such that $x^2 = -9$.*" The symbolic form is then

$$\sim \exists x \in \mathbb{R}\ (x^2 = -9).$$

$\square$

Quantifiers are often nested in mathematical statements. The order of the quantifiers can make a difference when determining whether a quantified predicate is true or false. For example, let $p(x, y)$ denote the predicate,

"$x + y = 9$," and let the domain of discourse for both x and y be the set of all real numbers. Then the proposition

$$\forall x \, \exists y \, p(x, y)$$

has the meaning,

For all x there is a y such that $x + y = 9$.

This is a true proposition, since given any x, we can let $y = 9 - x$, and then it follows that $x + y = 9$. The proposition

$$\exists y \, \forall x \, p(x, y)$$

means,

There is some y such that for every x, $x + y = 9$.

This proposition is false because there is no y such that for every real number x, $x + y = 9$. The difference between $\forall x \, \exists y \, p(x, y)$ and $\exists y \, \forall x \, p(x, y)$ is that in the former proposition, the choice of y can depend on x, whereas in the latter proposition, there must be some y which works for every x.

For any predicate $p(x, y)$, the propositions $\exists x \, \exists y \, p(x, y)$ and $\exists y \, \exists x \, p(x, y)$ are logically equivalent. Sometimes the shorthand notation

$$\exists x, y \; p(x, y)$$

is used to denote $\exists x \, \exists y \, p(x, y)$. Likewise,

$$\forall x, y \; p(x, y)$$

is sometimes used instead of the logically equivalent propositions $\forall x \, \forall y \, p(x, y)$ and $\forall y \, \forall x \, p(x, y)$.

Example 1.3.5. Let $p(x, y)$ denote the predicate, "*x made an A in y*," where the domain of discourse for x is the set of students listed in the table below and the domain of discourse for y is the set of classes in the same table.

	Calculus	Biology	Physics	English	History
Chris	A	A	B	A	B
Patty	A	C	B	C	C
Eddie	D	A	A	D	B
Bobby	A	A	A	A	A
Charley	B	C	F	B	A

According to the table, we can make the following conclusions:

- $\exists x, y\, p(x, y)$ is true because there is a student that made an A in some class. (There is an A somewhere in the table.)
- $\exists x\, \forall y\, p(x, y)$ is true because there is a student who made all A's. (There is a horizontal row consisting of only A's.)
- $\forall x\, \exists y\, p(x, y)$ is true because every student made at least one A. (There is at least one A in each horizontal row.)
- $\forall x, y\, p(x, y)$ is false because it is not true that every student made straight A's. (There is a grade other than A somewhere in the table.)
- $\exists y\, \forall x\, p(x, y)$ is false because there is no class in which every student made an A. (There is no vertical column consisting of only A's.)
- $\forall y\, \exists x\, p(x, y)$ is true because in each class there is a student who made an A. (There is at least one A in each vertical column.)

$\square$

Example 1.3.6. In calculus, the definition of a limit is as follows:

$$\lim_{x \to a} f(x) = L$$

means that for every $\epsilon > 0$ there is a corresponding $\delta > 0$ such that $|f(x) - L| < \epsilon$ whenever $0 < |x - a| < \delta$; that is, if

$$\forall \epsilon > 0\ \exists \delta > 0\ \forall x\ (0 < |x - a| < \delta \to |f(x) - L| < \epsilon).$$

Notice that the quantification $\forall x$ is assumed from the context (see Remark 1.3.3). $\square$

Consider the proposition, "*All integers greater than 30 are prime.*" In symbolic form, this proposition can be written as $\forall n\, p(n)$ where the domain of discourse is the set of all integers greater than 30 and where $p(n)$ denotes the predicate "*n is prime.*" Clearly this proposition is false, but how would you convince another person that it is false? A little thought might lead you to the argument that the integer 32 is greater than 30, but 32 is not prime, so the proposition cannot be true. What makes the proposition $\forall n\, p(n)$ false is that there is an integer n in the domain of discourse such that $p(n)$ is false. Hence, the negation of $\forall n\, p(n)$ is logically equivalent to $\exists n \sim p(n)$; that is,

$$\sim \forall n\, p(n) \Leftrightarrow \exists n \sim p(n).$$

This is true in general; if $p(x)$ is any predicate, then

$$\sim \forall x\, p(x) \Leftrightarrow \exists x \sim p(x).$$

This logical equivalence is similar to DeMorgan's Law (Theorem 1.2.3 (a)). Indeed, if the domain of discourse consists of only the values 0, 1 and 2, then

$$\sim \forall x \; p(x)$$

is equivalent to

$$\sim (p(0) \wedge p(1) \wedge p(2)).$$

By DeMorgan's Law, this is equivalent to

$$\sim p(0) \vee \sim p(1) \vee \sim p(2),$$

which in turn is equivalent to

$$\exists x \sim p(x).$$

The propositions, *"All integers greater than 30 are prime"* and, *"All integers greater than 30 are not prime"* are both false, so the negation of $\forall n\, p(n)$ is not logically equivalent to $\forall n \sim p(n)$. (Remember that a proposition and its negation must have opposite truth values.)

In English, the propositions, *"All integers greater than 30 are not prime"* and, *"Not all integers greater than 30 are prime"* have different meanings; the former is false because 31 is an integer greater than 30 which is prime, but the latter is true because 32 is an integer greater than 30 which is not prime.

The negation of an existentially quantified predicate is a universally quantified negation of the predicate. For example, consider the existentially quantified predicate, *"There is an integer n such that $n^2 < 0$."* This is false because for every integer n, $n^2 \geq 0$. In general,

$$\sim \exists x\, p(x) \Leftrightarrow \forall x \sim p(x).$$

Note that the domain of discourse does not change when we rewrite the negation of a quantified predicate as a quantified negation. For example, we would rewrite $\sim \exists x \in \mathbb{R}\, (p(x))$ as $\forall x \in \mathbb{R}\, (\sim p(x))$.

The negation of a multi-quantified predicate is logically equivalent to a multi-quantified negation, where we switch all the universal quantifiers to existential quantifiers and vice-versa. For example, $\sim \forall x \, \exists y \, \exists z \, \forall n \, p(x, y, z, n)$ is logically equivalent to $\exists x \, \forall y \, \forall z \, \exists n \sim p(x, y, z, n)$. The domain of discourse does not change.

Exercises

1.3.1. Determine whether each of the following propositions is true or false.
(a) $\exists x \in \mathbb{R} \, (x^2 + 4x + 4 = 0)$
(b) $\exists x \in \mathbb{R} \, (x^2 + 4x + 1 = 0)$
(c) $\exists x \in \mathbb{R} \, (x^2 + 4x + 5 = 0)$
(d) $\exists! x \in \mathbb{R} \, (x^2 + 4x + 4 = 0)$
(e) $\exists! x \in \mathbb{R} \, (x^2 + 4x + 1 = 0)$
(f) $\exists! x \in \mathbb{R} \, (x^2 + 4x + 5 = 0)$

1.3.2. Determine whether each of the following propositions is true or false.
(a) $\exists x \in \mathbb{R} \, (\, x^2 + 2x + 1 = (x + 1)^2 \,)$
(b) $\exists! x \in \mathbb{R} \, (\, x^2 + 2x + 1 = (x + 1)^2 \,)$
(c) $\forall x \in \mathbb{R} \, (\, x^2 + 2x + 1 = (x + 1)^2 \,)$
(d) $\exists x \in \mathbb{R} \, (\, x^2 + 2x + 1 = (x + 2)^2 \,)$
(e) $\exists x \in \mathbb{R} \, (\, x^2 + 2x + 1 = -(x + 2)^2 \,)$
(f) $\forall x \in \mathbb{R} \, (\, x^2 + 2x + 1 = -(x + 2)^2 \,)$

1.3.3. Let $p(x)$ denote the predicate, "$4x + 3 = 0$." Determine whether each of the following propositions is true or false.
(a) $\exists x \in \mathbb{C} \ p(x)$
(b) $\exists x \in \mathbb{R} \ p(x)$
(c) $\exists x \in \mathbb{R}^+ \ p(x)$
(d) $\exists x \in \mathbb{Q} \ p(x)$
(e) $\exists x \in \mathbb{Q}^+ \ p(x)$
(f) $\exists x \in \mathbb{Z} \ p(x)$
(g) $\exists x \in \mathbb{N} \ p(x)$

1.3.4. Write each of the following in symbolic form.
(a) All integers are positive.
(b) No integers are positive.
(c) Some integers are positive.
(d) Some integers are not positive.

(e) Every integer is positive.

(f) There is at least one integer which is positive.

(g) There is one and only one integer which is positive.

1.3.5. Determine whether each of the following propositions is true or false. Assume that the domain of discourse for both x and y is the set of all real numbers.

(a) $\exists x \, \forall y \, (x \le y)$

(b) $\exists x, y \, (x \le y)$

(c) $\forall y \, \exists x \, (x \le y)$

(d) $\forall x, y \, (x \le y)$

(e) $\exists y \, \forall x \, (x \le y)$

(f) $\forall x \, \exists y \, (x \le y)$

1.3.6. Determine whether each of the following propositions is true or false. Assume that the domain of discourse for both x and y is the set of all natural numbers.

(a) $\exists x \, \forall y \, (x \le y)$

(b) $\exists x, y \, (x \le y)$

(c) $\forall y \, \exists x \, (x \le y)$

(d) $\forall x, y \, (x \le y)$

(e) $\exists y \, \forall x \, (x \le y)$

(f) $\forall x \, \exists y \, (x \le y)$

1.3.7. Determine whether each of the following propositions is true or false. Assume that the domain of discourse for both x and y is the set of all real numbers.

(a) $\exists x \, \forall y \, (y = x^2)$

(b) $\exists x, y \, (y = x^2)$

(c) $\forall y \, \exists x \, (y = x^2)$

(d) $\forall x, y \, (y = x^2)$

(e) $\exists y \, \forall x \, (y = x^2)$

(f) $\forall x \, \exists y \, (y = x^2)$

1.3.8. Determine whether each of the following propositions is true or false. Assume that the domain of discourse for both x and y is the set of all real numbers.

(a) $\exists x \, \forall y \, (y \le x^2)$

(b) $\exists x, y \, (y \le x^2)$

(c) $\forall y \, \exists x \, (y \leq x^2)$
(d) $\forall x, y \, (y \leq x^2)$
(e) $\exists y \, \forall x \, (y \leq x^2)$
(f) $\forall x \, \exists y \, (y \leq x^2)$

1.3.9. Let $p(x, y)$ denote the predicate, "*x loves y.*" Let the domain of discourse be the set of all people. Write each of the following propositions in symbolic form.
(a) Everybody loves somebody.
(b) Somebody loves everybody.
(c) Only one person loves everybody.
(d) There is a person whom everybody loves.
(e) Everybody has somebody who loves them.

1.3.10. Write the negation of each of the following quantified predicates.
(a) $\exists x \in \mathbb{R} \, (x > 3)$
(b) $\forall y \in \mathbb{R} \, (y^3 \leq 7)$
(c) $\forall x \, \exists y \, (x + y = 7)$
(d) $\forall x, y \, (x + y > 7)$
(e) $\exists x, y \, (x = y - 4)$
(f) $\forall x \in \mathbb{R} \, \exists y \in \mathbb{R} \, (y = x^2)$
(g) $\exists x \, \forall z \, \exists y \, (z = x + y)$

1.3.11. Write the negation of each of the following. In your answers, do not use the conditional connective ($\rightarrow$). Proposition 1.2.6 might help here.
(a) $\forall x \in \mathbb{R} \, ((x > 5) \rightarrow (x^2 > 25))$
(b) $\forall n \in \mathbb{Z} \, \exists m \in \mathbb{Z} \, ((n^2 = m^2) \rightarrow (n = m))$
(c) $\forall n \in \mathbb{Z} \, \forall m \in \mathbb{Z} \, ((n^2 = m^2) \rightarrow (n = m))$
(d) $(\exists x \in \mathbb{R} \, x^2 = 5) \rightarrow (\exists x \in \mathbb{C} \, x^2 = 5)$

1.3.12. Use Example 1.3.6 and Proposition 1.2.6 to write what it means for

$$\lim_{x \to a} f(x) \neq L.$$

1.3.13. Write the negation of the proposition $\exists! x \, \, x^2 < 2$ without just placing the symbol $\sim$ in front of it.

Chapter 2

Proof techniques

Mathematics seems to endow one with something like a new sense.

-Charles Darwin, naturalist and evolutionary biologist (1809-1882)

2.1   Making conjectures

Mathematics is about the discovery and explanation of patterns or consistent traits that appear in both abstract and concrete settings. These patterns are often geometric, algebraic or numerical in form, but even so, mathematics is often considered an art. Just as you can gaze at a painting and suddenly see something you hadn't noticed before, you can study a collection of numbers or a complicated geometric figure and discover a pattern. You can also study a new object and suddenly notice that the new object has something in common with another object you had already studied before. Mathematicians often become excited when they notice how two seemingly unrelated ideas are actually very similar.

How to discover a pattern is difficult to learn, but discoveries usually just happen after some time is spent studying the form. This can make the process somewhat frustrating, but the more time you spend to get to a discovery, the more exciting the discovery will be.

When a pattern is first noticed, the idea is usually expressed as a conjecture. A **conjecture** is a proposition which someone proposes to be true, but no one has yet proven to be true. In mathematics, a **proof** is a sequence of statements that form a deductive argument and demonstrates that a proposition is true. The statements in a proof usually consist of accepted assumptions, accepted definitions of the terms involved in the proposition, and known facts which have already been proven. If a proof of the conjecture is found, the conjecture becomes a theorem. Thus, a **theorem** is a proposition for which there is a logically valid proof. The term *theorem* should not be confused with the term *theory*; a *theory* is a speculation or an unproved assumption which may or may not be factual. A *theorem* is usually considered to be an indisputable fact. The term *theory* is also used to describe a body of general principles to explain phenomena, as in the *theory of relativity*.

Mathematics is unique in that deductive proofs are used to verify statements. In other branches of science, statements are verified through experimentation and empirical evidence. Certainly experimentation and investigation play a large role in mathematics, but they do not distinguish mathematics from other fields.

A **lemma** is a theorem used to simplify the proofs of other theorems. If the proof of a theorem is excessively long, several lemmas might be constructed to break up the theorem into parts. A **corollary** is a theorem which follows easily from another theorem. Outside the realm of symbolic logic, the term **proposition** is used to describe a theorem which is relatively simple (as compared to more general or profound theorems). An **axiom** is a basic fact that is assumed to be true and has no proof. The term **postulate** is synonymous with *axiom*, but often a *postulate* pertains specifically to geometry.

Before we get to the study of proofs, we should get a little practice with the process of making conjectures. Often conjectures are the result of inductive reasoning, as in the following example.

Example 2.1.1. Compute several values of $f(n) = n^2 - n + 41$ for nonnegative integers n, and make a conjecture about the results.

Solution. We can plug in the integers $n = 0, 1, 2, 3, \ldots$ to obtain the following outputs: $41, 43, 47, 53, 61, 71, 83, \ldots$. What do all of these outputs have in common? Many answers are possible. A few conjectures are possible:

1. For every nonnegative integer n, $f(n)$ is odd.

2. For every nonnegative integer n, the last digit of $f(n)$ is a 1, 3, or 7.

3. For every nonnegative integer n, $f(n)$ is prime.

Many other conjectures are possible also. Which of the above conjectures do you think are actually true? □

Exercises

2.1.1. Show that the third conjecture in Example 2.1.1 is false. (You might want to compare this to Example 1.2.7.)

2.1.2. Compute 11^2, 11^3, and 11^4. Then make a conjecture. Test your conjecture.

2.1.3. Compute some values of the expression

$$\frac{1}{\sqrt{5}} \left(\left(\frac{1 + \sqrt{5}}{2} \right)^n - \left(\frac{1 - \sqrt{5}}{2} \right)^n \right)$$

for some nonnegative integers n. Make a conjecture about the results.

2.1.4. Pick any positive integer n. If n is even, divide it by 2. If it is odd, multiply it by 3 and then add 1. Then repeat this process with the result; if the result is even, divide it by 2, and if it is odd, multiply it by 3 and add 1. Repeat this process until you can predict the next few numbers in this sequence.

Now do the same but starting with different positive integers n. Make a conjecture about the results.

2.1.5. Let $f(x)$ be the function given by

$$f(x) = x^{1/\ln x}.$$

Compute some values of $f(x)$ for positive numbers x. Make a conjecture about this function.

2.1.6. For various values of the positive integer n, compute the sum of the first n positive odd integers. Make a conjecture about the result.

2.1.7. Let n be a nonnegative integer. Let d denote the $(n+1)$-digit integer whose first n digits are all 3 and whose last digit is 4. (For example, when $n = 2$, $d = 334$.) Make a conjecture about the digits in the integer d^2.

2.1.8. Pick a real number r, and use a calculator to approximate the value of $\cos r$. Then approximate the cosine of the result. Then approximate the cosine of that, and continue until you see a pattern. Make a conjecture about what happens for different values of r.

2.1.9. Draw any number of dots on a sheet of paper. Connect the dots with as many non-intersecting line segments as you like, but make sure every dot is connected to every other dot via at least one sequence of line segments. The line segments separate the paper into regions; one region is outside, and perhaps several regions are inside surrounded by line segments. Count the numbers of dots d, line segments ℓ, and regions r, and compute $d - \ell + r$. Repeat this experiment and make a conjecture about the value of $d - \ell + r$.

2.1.10. Compute the first few derivatives of $f(x) = xe^x$. Make a conjecture about what the nth derivative of $f(x)$ is.

2.1.11. A *regular* polygon is a polygon in which all sides have the same length and all interior angles have the same measure. A *tiling* of the plane is a covering of the plane using polygons which do not overlap and do not leave any part of the plane uncovered. Make a conjecture about which regular polygons can be used to make a tiling of the plane. (Use only one kind of polygon for each tiling.)

2.1.12. Find the last digit of the integer $2^{2^{2018}}$.

2.2   Conditionals and contradictions

Recall from Section 1.2 that deductive reasoning is the kind of reasoning in which conclusions are based upon accepted assumptions and logically valid arguments. In that section we used truth tables to determine whether arguments were logically valid. This is a reliable yet tedious method, and we would like to get away from using truth tables and read and write English arguments rather than symbolic ones. Constructing truth tables and studying symbolic logic helps people to develop a sense of what kinds of arguments are valid. Most arguments follow one of only a few common forms, and it is our goal in this chapter to study these common forms.

An understanding of basic truth tables is important when reading or constructing a proof of a proposition. Since a large proportion of theorems in mathematics are conditional propositions, an understanding of the truth table for $p \rightarrow q$ is necessary before an understanding of a proof for a conditional proposition can develop.

Since $p \rightarrow q$ is true in both cases in which p is false, if we want to prove that $p \rightarrow q$ is true, we need only consider the case when p is true. So in order to begin a proof of the conditional proposition $p \rightarrow q$, we might start with one of the statements, "*Suppose p is true,*" "*Assume p is true,*" "*Suppose p,*" or "*Assume p.*" Then in order to finish the proof of $p \rightarrow q$, we need to deduce (perhaps from definitions, other theorems, or axioms) that q is also true. (In order for $p \rightarrow q$ to be true in the case where p is true, it is necessary that q is also true. This follows from the truth table for $p \rightarrow q$.) A **direct proof** of the conditional proposition $p \rightarrow q$ consists of supposing that p is true and deducing that q must also be true.

An example will illustrate the direct proof method, but it will be helpful to first give a simple definition.

Definition 2.2.1. An integer n is **even** if there exists an integer k such that $n = 2k$.

Example 2.2.2. Give a direct proof of the proposition, *If n is even, then n^2 is even.*

Solution. The first step in a direct proof of a conditional proposition is to assume the hypothesis:

> *Suppose n is even.*

Then we want to deduce that n^2 must be even too. We might consider what our hypothesis means. We can write this down as a second sentence:

> *Then by the definition of even, there exists an integer k such that $n = 2k$.*

Since we want to prove something about n^2, we might next square both sides of the equation $n = 2k$:

> *Then $n^2 = (2k)^2$.*

Now that we have a formula for n^2, we consider whether the formula tells us that n^2 is even. Since $(2k)^2 = 2 \cdot (2k^2)$ and $(2k^2)$ is an integer, it does. The fact that $2k^2$ is actually an integer, as opposed to any real number, is important;

for example, $7 = 2 \cdot (7/2)$, but 7 is not even since $(7/2)$ is not an integer. The fact that $2k^2$ is an integer follows from the fact that when we multiply integers together, the product must also be an integer. Because the product of integers is always an integer, we say that the integers are **closed under multiplication**. The integers are also **closed under addition** and **closed under subtraction** because the sum and the difference of two integers must be an integer.

We finish the proof as follows:

> *Since $(2k)^2 = 2 \cdot (2k^2)$, we get that $n^2 = 2 \cdot (2k^2)$. Since $(2k^2)$ is an integer, n^2 is even by definition.*

Many people choose to point out the end of a proof by writing **q.e.d.** after the proof, which stands for **quod erat demonstratum**, Latin for *that which has been demonstrated*. The finished proof can look like this:

> *Suppose n is even. Then by the definition of even, there exists an integer k such that $n = 2k$. Then $n^2 = (2k)^2$. Since $(2k)^2 = 2 \cdot (2k^2)$, we get that $n^2 = 2 \cdot (2k^2)$. Since $(2k^2)$ is an integer, n^2 is even by definition.* **q.e.d.**

$\square$

Example 2.2.3. Construct an appropriate first sentence in a direct proof for the following theorem from Real Analysis, known as *Fatou's Lemma*:

> If $(X, \mathcal{M}, \mu)$ is a measure space and $\{f_n\}$ is a sequence of non-negative extended real-valued measurable functions defined on X, then
> $$\int_X \liminf f_n \, d\mu \leq \liminf \int_X f_n \, d\mu.$$

Solution. At first glance, the problem might seem intimidating because of all of the terminology that you probably do not recognize. However, we do not need to understand the theorem to just construct a first line in a direct proof. An appropriate first sentence would be,

> *Suppose $(X, \mathcal{M}, \mu)$ is a measure space and $\{f_n\}$ is a sequence of nonnegative extended real-valued measurable functions defined on X.*

We are only assuming that the hypothesis is true. (To finish the rest of the proof is another matter completely.) □

The direct proof is not the only method of proving a conditional proposition. In Example 1.2.1 we saw that a conditional proposition $p \to q$ is logically equivalent to its contrapositive $\sim q \to \sim p$. Hence, to prove $p \to q$, it suffices to prove $\sim q \to \sim p$. The **proof by contrapositive** of a conditional proposition $p \to q$ is a direct proof of $\sim q \to \sim p$. Sometimes it is easier to construct a proof by contrapositive than a direct proof. Before we give an example, we need a definition and a theorem.

Definition 2.2.4. An integer n is **odd** if there exists an integer k such that $n = 2k + 1$.

Theorem 2.2.5. *Every integer is either even or odd, and no integer is both even and odd.*

A proof of Theorem 2.2.5 would lead us away from our immediate point, so we will not include it here. We will prove the second part of this theorem in Example 2.2.7 momentarily.

Example 2.2.6. Prove: *If n^2 is even, then n is even.*

Solution. We might first attempt to construct a direct proof, and see where that leads:

> *Suppose n^2 is even. Then by the definition of even, there is some integer k such that $n^2 = 2k$.*

Next, we would want to deduce that n is even. Since $n^2 = 2k$, we can solve for n to get $n = \pm\sqrt{2k}$. However, it is not clear how to proceed from there; it is not obvious that there is an integer ℓ such that $\pm\sqrt{2k} = 2\ell$.

A proof by contrapositive is easier. A proof by contrapositive is a direct proof of the proposition, *If n is not even, then n^2 is not even*:

> *Suppose n is not even. Then by Theorem 2.2.5, n is odd. By the definition of odd, there exists an integer k such that $n = 2k + 1$. Hence, $n^2 = (2k + 1)^2$. Since*
>
> $$(2k + 1)^2 = 4k^2 + 4k + 1 = 2(2k^2 + 2k) + 1$$
>
> *and $(2k^2 + 2k)$ is an integer, n^2 is odd by the definition of odd.*
>
> **q.e.d.**

One method of proving a (not necessarily conditional) proposition p is to suppose that p is not true and deduce that some other proposition c is true, where c is a contradiction. A **contradiction** is a proposition which is always false. If it is proven that the conditional proposition $(\sim p) \to c$ is true, and we know that c is always false, then the hypothesis $\sim p$ must be false also. This makes p true. This proof method is known as the **proof by contradiction**. Often the contradiction is of the form $q \wedge \sim q$ (like n is even and n is not even, for example), but an obviously false proposition like "$0 = 1$" would work as a contradiction as well.

Example 2.2.7. Prove that no integer is both even and odd.

Solution. We will prove the statement by contradiction, so we start by supposing that the statement is false. The negation of the statement is that some integer is both even and odd. We explain what that statement means, and eventually end with a contradiction:

> *Suppose there is an integer which is both even and odd. Call this integer M. By the definition of even, there must be an integer k such that $M = 2k$. By the definition of odd, there must be an integer ℓ such that $M = 2\ell + 1$. Hence, $2k = 2\ell + 1$, and dividing both sides of this equation by 2 yields the equality $k = \ell + 1/2$. Therefore, $k - \ell = 1/2$. But since k and ℓ are integers, $k - \ell$ must also be an integer; so $k - \ell$ cannot equal $1/2$. By contradiction, there can be no integer which is both even and odd.* **q.e.d.**

$\square$

To give another example of a proof by contradiction, we first need a couple of definitions.

Definition 2.2.8. The integer a **divides** the integer b if there exists an integer c such that $ac = b$. If a divides b, then we write $a \mid b$ and we say that a is a **divisor** or a **factor** of b, and that b is **divisible** by a. If a divides b, we also say that b is a **multiple** of a. The integer a is a **common divisor** or a **common factor** of the integers n and m if $a \mid n$ and $a \mid m$.

Example 2.2.9. Prove that $\sqrt{2}$ is irrational.

Solution. We will prove the proposition by contradiction:

Suppose $\sqrt{2}$ is not irrational. Then $\sqrt{2}$ is rational. By the definition of rational number (see Page 23), $\sqrt{2}$ can be expressed as n/m, where $m \neq 0$ and where n and m are integers. By dividing n and m by their greatest common divisor, if necessary, we can assume that n and m have no common divisors other than ± 1. Squaring both sides of the equation $\sqrt{2} = n/m$ yields $2 = n^2/m^2$, and multiplying by m^2 yields $2m^2 = n^2$. Hence, n^2 is even. By the proposition proven in Example 2.2.6, n must be even. Hence, there must be an integer ℓ such that $n = 2\ell$. Then $n^2 = 4\ell^2$, so $2m^2 = 4\ell^2$. Dividing by 2 yields the equation $m^2 = 2\ell^2$, and it follows that m^2 is even. Using Example 2.2.6 again, we get that m is even. Since n and m are both even, 2 is a common divisor of n and m. This is a contradiction to the assertion that n and m have no common divisors other than ± 1. **q.e.d.**

□

Example 2.2.10. Prove that $f(x) = x^3 + 5x^2 + 2x + 1$ has no positive real roots.

Solution. We give a proof by contradiction. Suppose that $f(x) = x^3 + 5x^2 + 2x + 1$ does have a positive real root r. Then $r > 0$ and $f(r) = 0$. But $f(r) = r^3 + 5r^2 + 2r + 1$, and since $r > 0$, $r^3 + 5r^2 + 2r + 1 > 1$. This contradicts the assertion that $f(r) = 0$, so our supposition that $f(x)$ had a positive real root must have been false. **q.e.d.**

Example 2.2.11. In Euclid's classic *Elements*, he uses a proof by contradiction to prove that there are infinitely many primes. The proof proceeds as follows. Suppose that there are not infinitely many primes. Then there are only finitely many. List all of the primes: p_1, p_2, ..., p_n. Then since the integer $p = p_1 p_2 \cdots p_n + 1$ is greater than each of the primes on the list, p is not on the list and it is therefore not prime. It follows that some prime p_i divides p. But p_i divides the product $p_1 p_2 \cdots p_n$, so by Exercise 2.2.6, p_i must divide the difference $p - p_1 p_2 \cdots p_n = 1$. The conclusion that $p_i \,|\, 1$ is a contradiction, since no prime can divide 1. □

Note 2.2.12. In order to **prove the conditional statement $p \to q$ by contradiction**, we suppose that p is true and that q is false. Then we must find a contradiction which follows from this supposition.

The reason this works is that when we want to prove $p \to q$ by contradiction, the first step is to suppose that $p \to q$ is false. That is, we suppose that $\sim (p \to q)$ is true. By Proposition 1.2.6, $\sim (p \to q)$ is logically equivalent to $p \wedge \sim q$. Hence, $\sim (p \to q)$ is true exactly when p is true and q is false. Therefore, to suppose $p \to q$ is false is the same as to suppose p is true and q is false.   □

Example 2.2.13. Let n and m be integers. Prove that if n is odd and nm is even, then m is even.

Solution. The only way to immediately get useful expressions for n, m, and nm is by way of contradiction, so this is perhaps a good method to use.

> *By way of contradiction, suppose that n is odd, nm is even, and m is not even. Then n is odd and m is odd, so there are integers k and ℓ such that $n = 2k + 1$ and $m = 2\ell + 1$. Thus, $nm = (2k + 1)(2\ell + 1) = 2(2k\ell + k + \ell) + 1$. Since $(2k\ell + k + \ell)$ is an integer, nm must be odd. But we supposed that nm was even, so this is a contradiction.*   **q.e.d.**

□

Exercises

2.2.1. Explain what is wrong with the following proof of the proposition, *If n^2 is even, then n is even.*

> Suppose n^2 is even. Then by the definition of even, there exists an integer k such that $n^2 = (2k)^2$. Taking the square root of both sides, we get $n = \pm 2k = 2 \cdot (\pm k)$. Since k and $-k$ are integers, n is even by the definition of even.

2.2.2. Which of the following is an appropriate first step in a proof of the proposition, *If $f(x)$ is differentiable, then $f(x)$ is continuous?* Explain your answers.
(a) Suppose $f(x)$ is continuous.
(b) Suppose $f(x)$ is not continuous.
(c) Suppose $f(x)$ is differentiable.
(d) Suppose $f(x)$ is not differentiable.

(e) Suppose $f(x)$ is differentiable and $f(x)$ is continuous.
(f) Suppose $f(x)$ is differentiable and $f(x)$ is not continuous.
(g) Suppose if $f(x)$ is differentiable then $f(x)$ is continuous.
(h) Suppose if $f(x)$ is differentiable then $f(x)$ is not continuous.

2.2.3. What should you suppose in order to prove the proposition

> *If the number 0 is not an eigenvalue of A, then the determinant of A is not 0*

using:
(a) a direct proof?
(b) a proof by contrapositive?
(c) a proof by contradiction?

2.2.4. What should you suppose in order to prove the proposition

> *If either every left coset of N in G is also a right coset of N in G or aN = Na for all a in G, then left and right congruence modulo N are the same and $aNa^{-1} = N$ for all a in G*

using:
(a) a direct proof?
(b) a proof by contrapositive?
(c) a proof by contradiction?

2.2.5. Prove: *If $3n + 2$ is odd, then n is odd.*

2.2.6. Prove that if $n \mid a$ and $n \mid b$, then $n \mid (a + b)$ and $n \mid (a - b)$.

2.2.7. Which integers divide all integers? Which integers are divisible by all integers?

2.2.8. Prove: *If n and m are odd integers, then nm is odd.*

2.2.9. Let n and m be integers. Prove that if nm is even, then n is even or m is even.

2.2.10. Let n and m be integers. Prove that if $n + m$ is odd, then n is odd or m is odd.

2.2.11. Let n and m be integers. Prove that if n is odd and $n + m$ is even, then m is odd.

2.2.12. Prove that if the integers a and b are perfect squares, then the product ab is a perfect square.

2.2.13. Prove that $f(x) = 6x^5 + 7x^3 + 2x - 4$ has no negative real roots.

2.2.14. Give a proof by contradiction that there is no pair of integers x, y satisfying the equation $24x + 32y = 75$.

2.2.15. Use Proposition 1.2.4 to prove that for an integer n, n is odd or $n^2 + 1$ is odd.

2.3   Cases, biconditionals and quantifiers

In Exercise 1.1.10 you hopefully saw that the proposition

$$p \vee q \rightarrow r \leftrightarrow (p \rightarrow r) \wedge (q \rightarrow r)$$

is a tautology. Hence, the propositions $p \vee q \rightarrow r$ and $(p \rightarrow r) \wedge (q \rightarrow r)$ are logically equivalent, so to prove $p \vee q \rightarrow r$ we can prove $(p \rightarrow r) \wedge (q \rightarrow r)$ instead. To prove a conjunction, we prove each component of the conjunction; this divides the proof into two parts. This method of proving $p \vee q \rightarrow r$ by first proving $p \rightarrow r$ and then proving $q \rightarrow r$ is called a **proof by cases**.

Example 2.3.1. Let n and m be integers. Prove that if nm is odd, then n and m are both odd.

Solution. Let's first try a direct proof.

> *Suppose nm is odd. Then there exists an integer k such that $nm = 2k + 1$.*

Now we're stuck, since we don't have formulas for n and m independent of each other. We try a proof by contrapositive instead to make up for this. Discard our first attempt.

> *Suppose it is false that n and m are both odd. Then n is even or m is even.*

We now see that what we want to prove is, *If n is even or m is even, then nm is even.* This fits the form of $p \vee q \rightarrow r$, so we proceed now using a proof by cases. We must prove two conditional propositions: (1) if n is even, then nm is even; and (2) if m is even, then nm is even.

Case 1: Suppose n is even. Then there exists an integer k such that $n = 2k$. Hence, $nm = (2k)m = 2(km)$. Since (km) is an integer, nm is even.

Case 2: Suppose m is even. Then there exists an integer ℓ such that $m = 2\ell$. Hence, $nm = n(2\ell) = 2(n\ell)$. Since $(n\ell)$ is an integer, nm is even. **q.e.d.**

Note that we have used both the proof by contrapositive and the proof by cases methods in this proof.

Since the proofs in Case 1 and Case 2 above are almost identical, many authors will omit Case 2 and claim that "*by a similar argument, if m is even then nm is even*". Or even less precisely, they might omit Case 2 and use the words, "*Without loss of generality, suppose n is even*" at the beginning of the proof. This leaves the proof of the case where m is even up to the reader to complete. It is okay for an experienced proof-writer to use this rationale as long as they are absolutely sure that the cases are really so similar that the reader can effortlessly complete the other case by following the proof of the one case they do supply. But be warned that the use of this rationale is a very common source of errors, since some cases might seem more similar than they turn out to be. You should avoid omitting cases with the rationale *by a similar argument* or *without loss of generality* until you have a few years of experience with writing proofs. □

Example 2.3.2. Prove that if x is real then $|-3x| = 3|x|$.

Solution. Let's try a direct proof.

Suppose x is real.

Now it might not be clear at first where to go next, but recall the definition of the absolute value:

$$|x| = \begin{cases} x & \text{if } x \geq 0 \\ -x & \text{if } x < 0 \end{cases}$$

This definition gives us a clue to try a proof by cases.

Case 1: Suppose $x \geq 0$. Then $-3x \leq 0$, so $|-3x| = -(-3x) = 3x$. Also, since $x \geq 0$, $|x| = x$ and likewise $3|x| = 3x$. Therefore, $|-3x| = 3|x|$.

Case 2: Suppose $x < 0$. Then $-3x > 0$, so $|-3x| = -3x$. Also, since $x < 0$, $|x| = -x$ and $3|x| = -3x$. Thus, $|-3x| = 3|x|$.

q.e.d.

□

Note that a proof of a proposition of the form

$$p \to (q \vee r)$$

will not generally amount to a proof by cases; in Exercise 1.2.10, you saw that this proposition is logically equivalent to

$$(p \wedge \sim q) \to r,$$

and so a direct proof of this proposition can start by supposing the conjunction $p \wedge \sim q$, as opposed to a disjunction which might have proceeded in cases.

Next let's consider how to prove a general biconditional proposition

$$p \leftrightarrow q.$$

Recall that $p \leftrightarrow q$ was defined to have the same meaning as $(p \to q) \wedge (q \to p)$. Hence, we can split a proof of $p \leftrightarrow q$ into two parts and prove each part separately: (1) prove $p \to q$; and then also (2) prove $q \to p$.

Remark 2.3.3. Some mathematicians like to use the word **iff** as shorthand for *if and only if*.

Example 2.3.4. Prove: n is odd iff $7n + 3$ is even.

Solution. It is somewhat standard to break the two parts of a proof of a biconditional by beginning each part with one of the symbols $\Rightarrow$ or $\Leftarrow$:

($\Rightarrow$) *We give a direct proof of this implication. Suppose that n is odd. Then there exists an integer k such that $n = 2k + 1$. Hence, $7n + 3 = 14k + 10 = 2(7k + 5)$. Since $7k + 5$ is an integer, $7n + 3$ is even.*

($\Leftarrow$) *We will prove this implication by contrapositive. The contrapositive of the proposition, "If $7n + 3$ is even then n is odd" is the proposition, "If n is not odd then $7n + 3$ is not even." Suppose n is not odd. Then by Theorem 2.2.5, n is even. By definition, there exists an integer k such that $n = 2k$. Hence, $7n + 3 = 14k + 3 = 2(7k + 1) + 1$. Since $7k + 1$ is an integer, $7n + 3$ is odd. Therefore, $7n + 3$ is not even.* **q.e.d.**

Proofs of existentially quantified statements follow two types: **constructive** and **nonconstructive**. Constructive proofs consist of establishing either a specific value which makes the propositional function true or an algorithm or procedure for producing such a value. Nonconstructive proofs never establish such a specific value or even a method for producing one. The following are examples of constructive proofs.

Example 2.3.5. Prove that 12821 is odd.

Solution. Recall the definition of *odd*: an integer n is odd if there exists an integer k such that $n = 2k + 1$. This is an existentially quantified statement, and we can construct the specific value of k which makes the equality $12821 = 2k + 1$ true. Here is the proof:

> *Let $k = 6410$. Then k is an integer, and $12821 = 2k + 1$. Therefore, 12821 is odd.* **q.e.d.**

This proof is very short, as many constructive proofs are. However, the proof does not really show how we found the value of k to begin with, which can be frustrating for the reader. □

Example 2.3.6. Prove: $\exists x \in \mathbb{R} \ (x^2 + 2 > 10)$.

Solution. Here is a constructive proof:

> *Let $x = 5$. Then x is a real number, and $x^2 + 2 = 27 > 10$.* **q.e.d.**

Note that $x = 5$ is not the only choice we could have made. But the proof is logically correct. □

The proof of the Intermediate Value Theorem from calculus is nonconstructive. It is also quite difficult, and many textbooks on calculus omit the proof for this reason. But we can use the theorem to produce examples of other nonconstructive proofs.

Theorem 2.3.7 (The Intermediate Value Theorem). *Suppose $f(x)$ is continuous on the interval $[a, b]$ and let z be any number between $f(a)$ and $f(b)$. Then there exists a number c between a and b such that $f(c) = z$.*

Example 2.3.8. Prove: $\exists x \in \mathbb{R} \ (5x^6 + 7x^5 + 3x^2 + 2x - 3 = 0)$.

Solution. At first glance we might be tempted to try to solve the equation $5x^6 + 7x^5 + 3x^2 + 2x - 3 = 0$. But after a few attempts, you will probably realize just how difficult this might be. Here is a nonconstructive proof.

> Let $f(x) = 5x^6 + 7x^5 + 3x^2 + 2x - 3$. Then $f(0) = -3$ and
> $f(1) = 14$. Since $f(x)$ is continuous on the interval $[0, 1]$ and
> $z = 0$ is between -3 and 14, The Intermediate Value Theorem
> guarantees that there must be a number c between 0 and 1 such
> that $f(c) = 0$.   **q.e.d.**

$\square$

Constructive proofs are often used to disprove universally quantified statements. The specific value supplied in the constructive proof is called a **counterexample**. That is, because $\sim \forall x\ p(x)$ is logically equivalent to $\exists x \sim p(x)$, if we want to *disprove* $\forall x\ p(x)$, we can construct a specific value of x which makes $\sim p(x)$ true, and this value is called a counterexample.

Example 2.3.9. Prove or disprove: for all positive integers n, $n^2 - n + 11$ is prime.

Solution. You might recall this proposition from Example 1.2.7, where we found a counterexample. When $n = 11$, $n^2 - n + 11 = 121$, which is not prime. The value $n = 11$ is a counterexample.

> *Disproof:* Let $n = 11$. Then n is a positive integer, but $n^2 - n +$
> $11 = 121$ and 121 is not prime, since $121 = 11^2$.   **q.e.d.**

$\square$

To prove a statement of the form

$$\exists! x\ p(x),$$

it suffices to prove first that $\exists x\ p(x)$, and second that there do *not* exist two distinct values of x for which $p(x)$ is true. The second part is often achieved by contradiction: suppose that there are two distinct values x_1 and x_2 such that $p(x_1)$ and $p(x_2)$ are both true, and somehow deduce that $x_1 = x_2$, which will contradict the supposition that x_1 and x_2 are distinct.

To prove a universally quantified statement like $\forall x\ p(x)$, it is not enough to provide an example, or even several examples, when $p(x)$ is true. To prove $\forall x\ p(x)$, we might start with a statement like, *"Let x be given"* or *"Let x have any value* [in the domain of discourse]." Then we would try to prove that $p(x)$ is true, regardless of what value x takes.

Example 2.3.10. Prove: Every integer n is a divisor of 0.

Solution. The proposition is a universally quantified predicate. So we begin with the statement, *Let n be any integer.* Then we try to prove that n is a divisor of 0. Recall from Definition 2.2.8 that n is a divisor of 0 if there is an integer c such that $nc = 0$. Which value works for c?

> *Let n be any integer. Let c = 0. Then nc = 0, so by the definition*
> *of divisor, n is a divisor of* 0. **q.e.d.**

☐

Example 2.3.11. Use the definition given in Example 1.3.6 to prove that

$$\lim_{x \to 4} 3x - 2 = 10.$$

Solution. The definition of limit as it is applied in this case is:

$$\forall \epsilon > 0 \, \exists \delta > 0 \, \forall x \, (0 < |x - 4| < \delta \to |(3x - 2) - 10| < \epsilon).$$

We start the proof as we would start any proof of a universally quantified predicate:

> *Let $\epsilon > 0$ be given.*

The next step is to construct a specific value of δ which will make the predicate which follows true. Here we have to do some scratch work before we proceed. For a direct proof, we want to assume the hypothesis, and deduce that $|(3x - 2) - 10| < \epsilon$. But $|(3x - 2) - 10| < \epsilon$ is equivalent to $3|x - 4| < \epsilon$, or $|x - 4| < \epsilon/3$. Since our assumption will be that $0 < |x - 4| < \delta$, this gives us our value for δ: $\delta = \epsilon/3$. Here is the whole proof:

> *Let $\epsilon > 0$ be given. Let $\delta = \epsilon/3$. Let x be any real number.*
> *Suppose $0 < |x - 4| < \delta$. Then $|x - 4| < \epsilon/3$, so $|3x - 12| < \epsilon$.*
> *Hence, $|(3x - 2) - 10| < \epsilon$. This proves that $\lim_{x \to 4} 3x - 2 = 10$.*
> **q.e.d.**

☐

Exercises

2.3.1. Give a proof by cases that if n is any integer, then $n^2 - n + 11$ is odd.

2.3.2. Give a proof by cases that if x is real, then $|x| \geq x$.

2.3.3. Give a proof by cases that if x and y are real, then $|xy| = |x|\,|y|$.

2.3.4. Prove that n is even if and only if $3n^2$ is even.

2.3.5. Prove that n is odd if and only if $3n + 5$ is even.

2.3.6. Prove: For all integers n and m, $n + m$ is odd if and only if exactly one of n and m is odd.

2.3.7. Prove or disprove: $\exists x \in \mathbb{R}\ (3x = 5)$.

2.3.8. Prove or disprove: $\forall x \in \mathbb{R}\ (3x = 5)$.

2.3.9. Prove that 24593 is odd.

2.3.10. Prove: if n is even then $\exists! k \in \mathbb{Z}\ (n = 2k)$.

2.3.11. Prove or disprove: For every integer n, $1 \mid n$.

2.3.12. Prove: For all integers n, m and ℓ, if $n \mid m$, then $n \mid m\ell$.

2.3.13. Prove: For all integers n, m and ℓ, if $n \mid m$ and $n \mid \ell$ then $n^2 \mid m\ell$.

2.3.14. Prove: For all integers n, m and $\ell \neq 0$, $n \mid m$ if and only if $n\ell \mid m\ell$.

2.3.15. Prove: For all integers n, m and ℓ, if $n \mid m$ and $m \mid \ell$ then $n \mid \ell$.

2.3.16. Prove or disprove: For all integers n, m and ℓ, if $n \mid (m + \ell)$, then $n \mid m$ or $n \mid \ell$.

2.3.17. Prove or disprove: $\exists x \in \mathbb{R}\ (\,(x + 3)^2 = (x^2 + 9)\,)$.

2.3.18. Prove or disprove: $\forall x \in \mathbb{R}\ (\,(x + 3)^2 = (x^2 + 9)\,)$.

2.3.19. Prove that there is a real number x such that $x^4 + x - 9 = 0$.

2.3.20. Prove that there is a real number x such that $e^x = 9 - x^2$.

2.3.21. Prove that there is a real number x such that $x = \cos x$.

2.3.22. Let the domain of discourse for the variables n and m be the set of positive integers. Prove or disprove: $\forall n \, \exists m \, (m < n)$.

2.3.23. Let the domain of discourse for the variables n and m be the set of all integers. Prove or disprove: $\forall n \, \exists m \, (m < n)$.

2.3.24. Use the definition of limit given in Example 1.3.6 to prove that

$$\lim_{x \to 2} 4x - 9 = -1.$$

2.3.25. Use the definition of limit given in Example 1.3.6 to prove that

$$\lim_{x \to 1} 8 - 6x = 2.$$

2.4 Mathematical induction

In the last section we talked about how to prove a universally quantified statement like, $\forall x \, p(x)$ or $\forall n \, p(n)$. We began our proofs with general statements like, *Let x be any real number* or, *Let n be given*. In the particular case in which the domain of discourse is the set of all positive integers, there is a special method for proving the universally quantified statement, $\forall n \, p(n)$. In fact, given any integer m, the method can be applied to prove the more general statement, $\forall n \geq m \, (p(n))$.

Theorem 2.4.1 (The Principle of Mathematical Induction). *Let m be any integer. In order to prove the proposition*

> *For every integer $n \geq m$, $p(n)$*

it suffices to prove two simpler propositions:

(1) $p(m)$ is true; and

(2) for every integer $k \geq m$, the conditional proposition $p(k) \to p(k+1)$ is true.

In order for $p(n)$ to be true for every integer $n \geq m$, we need for $p(m)$ to be true, $p(m+1)$ to be true, $p(m+2)$ to be true, $p(m+3)$ to be true, and so forth. If we succeed in proving step (1) in the Principle of Mathematical Induction, then we can say that $p(m)$ is true. If we also succeed in proving step (2), then since we know $p(m)$ is true, $p(m+1)$ must also be true. By step (2) again, since $p(m+1)$ is true, $p(m+2)$ must be true. Following this logic forever will yield that $p(n)$ is true for all integers $n \geq m$:

$p(m)$ is true by (1)
$p(m) \rightarrow p(m+1)$ is true by (2)
$p(m+1)$ is true by modus ponens
$p(m+1) \rightarrow p(m+2)$ is true by (2)
$p(m+2)$ is true by modus ponens
$p(m+2) \rightarrow p(m+3)$ is true by (2)
$p(m+3)$ is true by modus ponens
$p(m+3) \rightarrow p(m+4)$ is true by (2)
$p(m+4)$ is true by modus ponens
$\vdots$

A proof by induction consists of two steps as shown in Theorem 2.4.1. The first step is called the **basis step**, where we prove that $p(m)$ is true. The second step is called the **inductive step**, where we prove the conditional proposition $p(k) \rightarrow p(k+1)$. We usually give a direct proof of $p(k) \rightarrow p(k+1)$, so that we start by supposing that $p(k)$ is true; this supposition is called the **inductive hypothesis**. By using the inductive hypothesis, we then need to deduce the conclusion $p(k+1)$. If we do this, then the Principle of Mathematical Induction asserts that $p(n)$ is true for all $n \geq m$.

In the induction hypothesis it is *not* assumed that $p(k)$ is true for all $k \geq m$; this is what induction is supposed to prove. In the induction hypothesis it is only assumed that $p(k)$ is true for *one particular* $k \geq m$. So we are not assuming that what we want to prove is true; we are proving it.

The Principle of Mathematical Induction is sometimes referred to as just "induction." The term *induction* should not be confused with *inductive reasoning*; these two ideas are very different.

Example 2.4.2. Prove that for every positive integer n, $5^n - 1$ is divisible by 4.

Solution. Before jumping into a proof of this proposition, it might be helpful to compute some examples so that we are sure we understand what it is we are trying to prove. We plug in some positive integers for n and check the proposition in those cases. Let's start with $n = 3$. When $n = 3$, $5^n - 1 = 124$. Since $124 = 4 \cdot 31$, 124 is divisible by 4. For another example, let's try $n = 7$. When $n = 7$, $5^n - 1 = 78124$. Since $78124 = 4 \cdot 19531$, 78124 is divisible by 4.

We could compute several more examples until we believe that the proposition is probably true, but as we saw in our discussion of inductive reasoning in Section 1.2, we should never be completely convinced by computing examples.

In this particular problem, the predicate $p(n)$ is the statement, "$5^n - 1$ is divisible by 4." Since we want to prove that $p(n)$ is true for all positive integers n, the value of m in the basis step is the smallest positive integer, $m = 1$. Here is what a proof might look like.

> *We use induction. For the basis step, we need to prove that $5^1 - 1$ is divisible by 4. Since $5^1 - 1 = 4 = 4 \cdot 1$, it is clear that $5^1 - 1$ is divisible by 4.*
>
> *Let k be any integer ≥ 1, and suppose that $5^k - 1$ is divisible by 4. We want to prove that $5^{k+1} - 1$ is also divisible by 4. Well, $5^{k+1} - 1 = 5 \cdot 5^k - 1 = (4 + 1) \cdot 5^k - 1 = 4 \cdot 5^k + 5^k - 1$. Since $4 \cdot 5^k$ is divisible by 4 and $5^k - 1$ is divisible by 4 (by the induction hypothesis), their sum $4 \cdot 5^k + 5^k - 1$ is also divisible by 4 (by Exercise 2.2.6). Hence, $5^{k+1} - 1$ is divisible by 4.*
>
> *By The Principle of Mathematical Induction, $5^n - 1$ is divisible by 4 for every positive integer n.*   **q.e.d.**

Notice how we started the proof with the sentence, *We use induction.* This is a good idea so that the reader knows what is coming next, and so the reader won't have to read through most of the proof before they realize what technique is being used. Many authors will not use a sentence like that unless the induction is subtle, as they will assume the reader is sophisticated enough to figure it out.

A completely different technique could have been used to prove that $5^n - 1$ is divisible by 4 for every positive integer n. Namely, $5^n - 1$ factors into

$$(5 - 1)(5^{n-1} + 5^{n-2} + \cdots + 5^1 + 1),$$

and given this factorization it is easy to see that $5^n - 1$ is divisible by 4. In general, $a^n - b^n$ factors into

$$(a - b)(a^{n-1} + a^{n-2}b + \cdots + a^1 b^{n-2} + b^{n-1}).$$

$\square$

The next example uses **summation notation**, which is the use of the Greek letter $\sum$ (capital sigma) to represent a sum. The notation

$$\sum_{n=k}^{M} f(n)$$

is used as shorthand for representing the sum

$$f(k) + f(k+1) + \cdots + f(M).$$

The variable n above is called the **index of summation**, and it does not affect the calculation. That is, we could use a different variable name and get the same result:

$$\sum_{n=k}^{M} f(n) = \sum_{i=k}^{M} f(i).$$

Several examples will illustrate the concept:

$$\sum_{n=2}^{4} 3n = 3 \cdot 2 + 3 \cdot 3 + 3 \cdot 4$$

$$\sum_{i=-1}^{2} i^2 = (-1)^2 + (0)^2 + 1^2 + 2^2$$

$$\sum_{k=3}^{6} x_k = x_3 + x_4 + x_5 + x_6$$

$$\sum_{m=4}^{5} (m^2 + m + 1) = (4^2 + 4 + 1) + (5^2 + 5 + 1)$$

$$\sum_{j=4}^{4} j^2 = (4^2)$$

In a similar fashion, the Greek letter $\prod$ (capital pi) represents a product:

$$\prod_{n=k}^{M} f(n) = f(k) \cdot f(k+1) \cdot \cdots \cdot f(M).$$

For example,

$$\prod_{n=2}^{4}(1-n^2) = (1-2^2) \cdot (1-3^2) \cdot (1-4^2) = -360.$$

Example 2.4.3. Prove that for every positive integer n,

$$\sum_{i=1}^{n}(2i-1) = n^2.$$

Solution. We first compute an example so we better understand the equality. This time let's use $n = 4$. Now

$$
\begin{aligned}
\sum_{i=1}^{4}(2i-1) &= (2 \cdot 1 - 1) + (2 \cdot 2 - 1) + (2 \cdot 3 - 1) + (2 \cdot 4 - 1) \\
&= 1 + 3 + 5 + 7 \\
&= 16
\end{aligned}
$$

and $4^2 = 16$, so the equality holds when $n = 4$. This is not a part of the proof, but it builds confidence in the result and verifies our understanding of the claim.

Proof: We use induction. The basis case is the case when $n = 1$. If $n = 1$ then $\sum_{i=1}^{n}(2i-1) = (2 \cdot 1 - 1) = 1$ and $n^2 = 1$, so the equality holds in this case.

Suppose

$$\sum_{i=1}^{k}(2i-1) = k^2.$$

We want to deduce from this supposition that

$$\sum_{i=1}^{k+1}(2i-1) = (k+1)^2.$$

Well, by taking the last term out of the summation we see that

$$\sum_{i=1}^{k+1}(2i-1) = \left[\sum_{i=1}^{k}(2i-1)\right] + (2(k+1)-1).$$

By the induction hypothesis,

$$\left[\sum_{i=1}^{k}(2i-1)\right] = k^2.$$

Hence,

$$\sum_{i=1}^{k+1}(2i-1) = k^2 + (2 \cdot (k+1) - 1) = k^2 + 2k + 1 = (k+1)^2.$$

q.e.d.

Notice that we just started with the left side of the equality

$$\sum_{i=1}^{k+1}(2i-1) = (k+1)^2$$

and simplified it using the inductive hypothesis until it looked precisely like the right hand side. □

Example 2.4.4. Use induction to prove that for every positive integer n,

$$\sum_{j=1}^{n}\left(\frac{1}{2}\right)^j = 1 - \left(\frac{1}{2}\right)^n.$$

Proof. We use induction. For the basis case, let $n = 1$. Then

$$\sum_{j=1}^{n}\left(\frac{1}{2}\right)^j = \sum_{j=1}^{1}\left(\frac{1}{2}\right)^j = \left(\frac{1}{2}\right)^1 = \frac{1}{2}.$$

Also, $1 - \left(\frac{1}{2}\right)^n = 1 - \left(\frac{1}{2}\right) = \frac{1}{2}$. Hence, when $n = 1$,

$$\sum_{j=1}^{n}\left(\frac{1}{2}\right)^j = 1 - \left(\frac{1}{2}\right)^n.$$

For the inductive step, let k be any positive integer and suppose

$$\sum_{j=1}^{k}\left(\frac{1}{2}\right)^j = 1 - \left(\frac{1}{2}\right)^k.$$

We want to show that

$$\sum_{j=1}^{k+1}\left(\frac{1}{2}\right)^{j}=1-\left(\frac{1}{2}\right)^{k+1}.$$

Well,

$$\begin{aligned}
\sum_{j=1}^{k+1}\left(\frac{1}{2}\right)^{j} &= \sum_{j=1}^{k}\left(\frac{1}{2}\right)^{j}+\left(\frac{1}{2}\right)^{k+1}\\
&= 1-\left(\frac{1}{2}\right)^{k}+\left(\frac{1}{2}\right)^{k+1}\\
&= 1-\left(\left(\frac{1}{2}\right)^{k}-\left(\frac{1}{2}\right)^{k}\cdot\frac{1}{2}\right)\\
&= 1-\left(\frac{1}{2}\right)^{k}\cdot\left(1-\frac{1}{2}\right)\\
&= 1-\left(\frac{1}{2}\right)^{k+1}.
\end{aligned}$$

q.e.d.

A couple of analogies might help one remember why the principle of induction works. First consider the activity of placing dominoes upright and close to one another, with the goal of knocking the first domino down and forcing each and every domino to be knocked down by the one next to it, one after another. This chain reaction is entertaining to watch, and people have held contests to see how long a chain of dominoes can be made with the constraint that every domino must eventually fall over. How can you be sure that all the dominoes will fall? We can be sure if two conditions are satisfied: (1) the first domino is actually pushed over; and (2) for every $k \geq 1$, if the kth domino falls over, then it will knock over the $(k+1)^{\text{st}}$ domino. Clearly this is the same idea as contained in the Principle of Mathematical Induction.

Second, consider climbing a ladder. You can climb the whole ladder, no matter how long it is, if the following two conditions are satisfied: (1) you can get onto the first rung; and (2) for every $k \geq 1$, if you can get onto the kth rung, then you can get onto the $(k+1)^{\text{st}}$ rung.

Another example might help. For the example, we use the notion of **n factorial**, denoted $n!$. For any integer $n \geq 1$, $n!$ is the product of all the integers from 1 to n. For example, $5! = 1 \cdot 2 \cdot 3 \cdot 4 \cdot 5$. Also, $0!$ is defined to be 1. Note that $5! = 5 \cdot (4!)$, and if $n \geq 1$, then $n! = n \cdot ((n-1)!)$.

Example 2.4.5. Prove that for every integer $n \geq 4$, $2^n < n!$.

Proof. We use induction. For the basis case, $n = 4$. When $n = 4$, $2^n = 16$ and $n! = 24$. Hence, when $n = 4$, $2^n < n!$.

Let k be any integer ≥ 4, and suppose that $2^k < k!$. We want to show that $2^{k+1} < (k+1)!$. Well, $2^{k+1} = 2 \cdot 2^k$. Since $k \geq 4$, $2 < (k+1)$. Hence, $2^{k+1} < (k+1) \cdot 2^k$. The induction hypothesis asserts that $2^k < k!$. By this hypothesis, $2^{k+1} < (k+1) \cdot k!$. Since $(k+1) \cdot k! = (k+1)!$, we get that $2^{k+1} < (k+1)!$.   **q.e.d.**

Another form of induction is occasionally used to prove statements of the form $\forall n > m$, $p(n)$:

Theorem 2.4.6 (Strong Form of Mathematical Induction). *In order to prove the proposition*

> *For every integer $n \geq m$, $p(n)$*

it suffices to prove two simpler propositions:
(1) $p(m)$ is true; and
(2) for every integer $k \geq m$, if $p(i)$ is true for all i satisfying $m \leq i \leq k$, then $p(k+1)$ is true.

The difference in the strong form is that the induction hypothesis is not just that $p(k)$ is true, but that $p(i)$ is true for all integers i satisfying $m \leq i \leq k$. Hence, we are assuming more in the strong form of induction. The strong form can be used when the proposition $p(k+1)$ does not naturally break up into a statement involving only $p(k)$, but rather, it breaks up into a statement involving $p(i)$ for perhaps several values of i between m and k.

The proof of the following theorem is a great example of an instance in which the strong form of induction can be used.

Theorem 2.4.7 (The Fundamental Theorem of Arithmetic). *Every integer $n \geq 2$ is either prime or can be expressed as a product of primes.*

The Fundamental Theorem of Arithmetic actually says more: except for the order of the primes that occur, the expression as a product of primes is unique. However, we omitted this part so that we could stay on track with our example of a proof which uses strong induction.

Proof. We use strong induction. The basis case is when $n = 2$. Since 2 is prime, the theorem holds in the basis case.

Let k be any integer ≥ 2, and suppose that for every integer i satisfying $2 \leq i \leq k$, either i is prime or i can be expressed as a product of primes. We must show that either $k + 1$ is prime or $k + 1$ can be expressed as a product of primes. Well, either $k + 1$ is prime or $k + 1$ is not prime. If $k + 1$ is prime, the conclusion holds. So let's consider the case when $k + 1$ is not prime. If $k + 1$ is not prime, then by the definition of prime, there must exist a pair of integers ℓ, m such that $2 \leq \ell \leq k$, $2 \leq m \leq k$, and $k + 1 = \ell \cdot m$. By the induction hypothesis, either ℓ is prime or ℓ can be expressed as a product of primes, and either m is prime or m can be expressed as a product of primes. Since $k + 1 = \ell \cdot m$, we can thus express $k + 1$ as a product of primes. **q.e.d.**

Exercises

2.4.1. Prove that for every integer $n \geq 1$, $11^n - 6$ is divisible by 5.

2.4.2. Prove that for every positive integer n, $7^n - 1$ is divisible by 6.

2.4.3. Prove that for every positive integer n, $7^n - 4$ is divisible by 3.

2.4.4. Prove that for every positive integer n, $n^3 - n$ is divisible by 3.

2.4.5. Prove that for every positive integer n, $n^5 - n$ is divisible by 5.

2.4.6. Prove that for every real number r with $r \neq 0$ and $r \neq 1$ and for every positive integer n,

$$r^0 + r^1 + r^2 + \cdots + r^n = \frac{1 - r^{n+1}}{1 - r}.$$

2.4.7. Prove that for every positive integer n,

$$\sum_{i=1}^{n} i = \frac{n(n+1)}{2}.$$

2.4.8. Prove that for every positive integer n,

$$\sum_{i=1}^{n} i^2 = \frac{n(n+1)(2n+1)}{6}.$$

2.4.9. Prove that for every positive integer n,

$$\sum_{i=1}^{n} i^3 = \frac{n^2(n+1)^2}{4}.$$

2.4.10. Prove that for every positive integer n,

$$\sum_{i=1}^{n} i(i!) = (n+1)! - 1.$$

2.4.11. Prove that for every integer $n \geq 3$,

$$\sum_{i=3}^{n} 3^i = \frac{3^{n+1} - 27}{2}.$$

2.4.12. Prove that for every integer $n \geq 2$,

$$\prod_{j=2}^{n} \left(1 - \frac{1}{j^2}\right) = \frac{n+1}{2n}.$$

2.4.13. Prove that for every positive integer n,

$$2^{n-1} \leq 2^n - 1.$$

2.4.14. Prove that for every integer $n \geq 7$, $3^n < n!$.

2.4.15. (a) Prove that for every integer $n \geq 3$,

$$2n + 1 < 2^n.$$

(b) Prove that for every integer $n \geq 4$,

$$n^2 \leq 2^n.$$

2.4.16. Use induction to prove that for every positive integer n, the nth derivative of x^n is $n!$.

2.4.17. Find a formula for the nth derivative of e^{2x}, and use induction to prove your answer.

2.4.18. Use induction to prove that for every positive integer n, the nth derivative of xe^x is $(x + n)e^x$.

2.4.19. The **Gamma Function** $\Gamma(z)$ is defined for $z > 0$ by the formula

$$\Gamma(z) = \int_0^\infty t^{z-1} e^{-t}\, dt.$$

Prove that for every integer $n \geq 0$,

$$\Gamma(n + 1) = n!$$

2.4.20. The **Fibonacci numbers** are the numbers in the sequence

$$0,\ 1,\ 1,\ 2,\ 3,\ 5,\ 8,\ 13,\ 21,\ 34,\ 55,\ 89,\ \ldots.$$

They are defined by the formulas

$$
\begin{aligned}
f_0 &= 0 \\
f_1 &= 1 \\
f_n &= f_{n-1} + f_{n-2}, \quad n \geq 2
\end{aligned}
$$

Prove that

$$f_n = \frac{1}{\sqrt{5}} \left(\left(\frac{1 + \sqrt{5}}{2} \right)^n - \left(\frac{1 - \sqrt{5}}{2} \right)^n \right)$$

for every integer $n \geq 0$. (The formula $f_n = f_{n-1} + f_{n-2}$ is called a **recursive formula** because it expresses the terms in a sequence as a formula involving the previous terms in the sequence.)

2.4.21. Use the Strong Form of Induction to prove that every positive integer is equal to a sum of distinct powers of 2.

2.4.22. Suppose we want to prove that $p(n)$ is true for all positive integers n, yet instead of the usual inductive step, we only see how to prove that for every positive integer k, $p(k) \rightarrow p(k + 2)$ is true. How can we change the basis step to conclude that $p(n)$ is true for all positive integers n?

2.4.23. The **Well-Ordering Principle** states that every nonempty subset of the set of positive integers has a smallest element. Use the Well-Ordering Principle to prove the Principle of Mathematical Induction.

2.4.24. Consider the proposition, *For all integers $n \geq 0$, $\cos n = 2^n$*. What is wrong with the following proof of this proposition?

> *We use induction. The basis case is the case $n = 0$. When $n = 0$, $\cos n = 1$ and $2^n = 1$, so the basis case holds.*
>
> *Let k be any integer ≥ 0, and suppose $\cos k = 2^k$. Then by substituting $k+1$ in for k, we get $\cos(k+1) = 2^{k+1}$. By induction, the proposition must be true for all $n \geq 0$.*

2.4.25. Consider the proposition, *All horses have the same color*. What is wrong with the following proof of this proposition?

> *The proposition can be rephrased as, "For every positive integer n, in any collection of n horses, every horse has the same color." We use induction to prove this. The basis case is the case $n = 1$. It is clear that in any collection consisting of 1 horse, every horse in this collection has the same color. So the basis case is satisfied. Suppose that $k \geq 1$, and that in any collection of k horses, every horse has the same color. We want to show that in any collection of $k+1$ horses, every horse has the same color. Suppose we have a collection of $k+1$ horses. Number these horses 1 through $k+1$. The horses 1 through k form a collection of k horses, so by the induction hypothesis every horse in this collection has the same color. The horses 2 through $k+1$ also form a collection of k horses, so by the induction hypothesis again, every horse in this collection has the same color. Since horses 1 through k all have the same color, and horses 2 through $k+1$ all have the same color, horses 1 through $k+1$ must all have the same color. Hence, all of the $k+1$ horses have the same color.*

Chapter 3

Set theory

By relieving the brain of all unnecessary work, a good notation sets it free to concentrate on more advanced problems, and, in effect, increases the mental power of the race.

-Alfred North Whitehead, philosopher and mathematician (1861-1947)

3.1 Set notation and Venn diagrams

A major goal of every branch of science is to classify the objects of study. Biology is the study of living organisms, and biologists classify living organisms into animals or plants, and they classify animals into their kingdom, phylum, class, order, family, genus, and species. Chemistry is the study of matter, and matter is classified by its molecular structure. The periodic table of elements is a table which classifies the basic building blocks of matter. The science of medicine is the study of the diagnosis, treatment and prevention of diseases and illnesses. A major goal of medicine is to classify diseases and illnesses so that they can be treated or prevented. Physics is the study of the interaction between matter and energy. In the field of physics one might classify the colors that occur in any rainbow.

In order to classify the objects of study, we can use the ideas of sets and elements. A **set** is a collection of items called **elements**. It is somewhat standard to use capital letters to denote sets, and to use lower-case letters to denote the elements in a set. If A is a set and x is an element of A, we write

$$x \in A.$$

If x is not an element of the set A, we write

$$x \notin A.$$

The elements of a set can be of any type; they can be colors, numbers, names, or even sets themselves. There are several different methods to specify a set. In the **descriptive method**, the set is described in words. For example, we might write, *Let S denote the set of colors in the rainbow.* In the **roster method**, the elements are specifically listed between braces:

Let $S = \{$red, orange, yellow, green, blue, indigo, violet$\}$.

Alternatively, we can use **set builder notation** to specify a set:

Let $S = \{x \mid x$ is a color in the rainbow$\}$

or

Let $S = \{x : x$ is a color in the rainbow$\}$.

The vertical line and the colon above are read aloud as "such that." The left brace ($\{$) is read aloud as "the set consisting of." So we would read the first expression aloud as, "Let S equal the set consisting of red, orange, yellow, green, blue, indigo and violet." The second and third expressions would be read aloud as, "Let S equal the set consisting of x such that x is a color in the rainbow."

The order in which the elements of a set are listed is irrelevant. For example, the expressions

$$A = \{1, 2, 6, 7\}$$

and

$$A = \{2, 7, 1, 6\}$$

have the same meaning. Note that $\{6\}$ is a set, but 6 is a number (not a set). Indeed, $6 \in A$ but $\{6\} \notin A$.

The number of elements in a set B is called the **cardinality** of the set and is denoted by $|B|$. With the examples above, $|S| = 7$ and $|A| = 4$. The **empty set** or **null set** is denoted by $\{\}$ or by $\emptyset$. However, $\{\emptyset\}$ is not the empty set, since there is one element in this set (namely, $\emptyset$). To give another example,

$$|\{\emptyset, \{1, 2, 3, 4, 5\}, \{7\}\}| = 3$$

because this set contains 3 elements. These 3 elements all happen to be sets themselves.

We will see in Exercise 3.1.15 that it does not make sense to define a set containing *everything*. Rather, we must restrict our attention to those items in context. The **universal set**, denoted by U, is the set of all elements under consideration. For example, U might be the set of all colors, or it might be the set of all real numbers.

We have already seen some of the more commonly used sets of numbers:

$$
\begin{array}{ll}
\mathbb{C} & \text{the set of all complex numbers} \\
\mathbb{R} & \text{the set of all real numbers} \\
\mathbb{R}^+ & \text{the set of positive real numbers} \\
\mathbb{Q} & \text{the set of all rational numbers} \\
\mathbb{Q}^+ & \text{the set of positive rational numbers} \\
\mathbb{Z} & \text{the set of all integers} \\
\mathbb{N} & \text{the set of all natural numbers}
\end{array}
$$

In general, the set A is called a **subset** of the set B if the following conditional proposition is true for every x in the universal set:

$$\text{if } x \in A \text{ then } x \in B.$$

We will use the notation $A \subseteq B$ to denote that A is a subset of B. Hence, in symbolic notation, $A \subseteq B$ means

$$\forall x \in U \ (x \in A \rightarrow x \in B).$$

The notation $A \supseteq B$ is used to denote that $B \subseteq A$; in this case, A is called a **superset** of B. (So if A is a superset of B, then B is a subset of A.) The two sets A and B are considered the same, and we write $A = B$, if $A \subseteq B$ and $B \subseteq A$. If A is a subset of B and the sets A and B are not the same set, then A is a **proper subset** of B. We will write $A \subsetneq B$ to denote that A is a proper subset of B. (Some textbooks use the notation $A \subset B$ to denote that A is a subset of B, and others use the same notation to denote that A is a proper subset of B. To avoid confusion, we will not use the symbol $\subset$.) The sets listed above are related in the following way:

$$\mathbb{N} \subsetneq \mathbb{Z} \subsetneq \mathbb{Q} \subsetneq \mathbb{R} \subsetneq \mathbb{C}.$$

Since for every x the hypothesis of the conditional proposition

$$\text{if } x \in \emptyset \text{ then } x \in B$$

must be false, the empty set is a subset of every set. It is a proper subset of every nonempty set.

The **power set** $\mathcal{P}(A)$ of a set A is the set of all subsets of A. For example, the power set of the set $S = \{1, 2, 3\}$ contains 8 elements:

$$\mathcal{P}(S) = \{\, \emptyset, \{1\}, \{2\}, \{3\}, \{1, 2\}, \{1, 3\}, \{2, 3\}, \{1, 2, 3\}\,\}.$$

The **union** of the sets A and B is defined to be

$$A \cup B = \{x \mid x \in A \text{ or } x \in B\}.$$

The **intersection** of A and B is

$$A \cap B = \{x \mid x \in A \text{ and } x \in B\}.$$

Hence, the intersection of two sets is the set of elements that they have in common, whereas the union is the set we get when we throw all the elements from the two sets into one. For example, let $A = \{1, 2, 3\}$ and $B = \{2, 3, 4, 5\}$. Then $A \cup B = \{1, 2, 3, 4, 5\}$ and $A \cap B = \{2, 3\}$.

The **set difference** $A - B$ is the set of all elements that are in A but not in B; that is,

$$A - B = \{x \mid x \in A \text{ but } x \notin B\}.$$

For example, $\{2, 4, 8, 9\} - \{1, 2, 3, 9\} = \{4, 8\}$. The **complement** of the set A is the set $U - A$; that is, it is the set containing everything under consideration that is not in A. The complement of A is denoted by $\overline{A}$ or A^c.

The sets A and B are **disjoint** if $A \cap B = \emptyset$; that is, they are disjoint if they have no elements in common. A collection of sets is **pairwise disjoint** if in every pair of distinct sets from the collection, the two sets are disjoint.

A **Venn diagram** is sometimes the best way to visualize sets. Such a diagram consists of a rectangle which represents the universal set and a collection of regions inside the rectangle which represent subsets of the universal set. Figure 3.1 shows a couple of Venn diagrams where the universal set is the set of integers from 1 to 9. Using this figure, one can visualize the following sets:

$$
\begin{aligned}
A \cup B &= \{3, 7, 5, 6, 9, 4\} \\
A \cap B &= \{5, 6\} \\
B - A &= \{9, 4\} \\
\overline{A} &= \{1, 2, 8, 4, 9\} \\
(C \cup D) \cap E &= \{6, 3, 1, 2\} \\
C \cup (D \cap E) &= \{6, 3, 1, 5, 2\}
\end{aligned}
$$

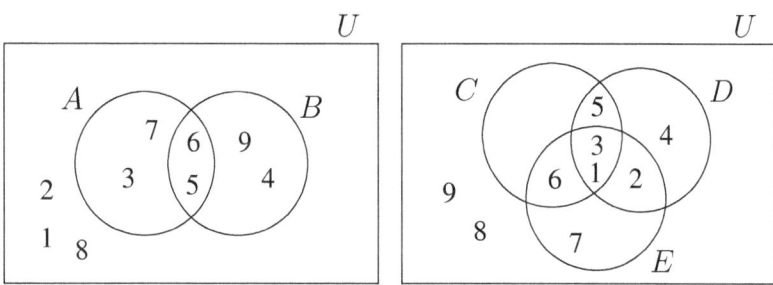

Figure 3.1: Venn diagrams for two and three sets

Note how the Venn diagram containing two sets divides the universal set into 4 pairwise disjoint regions: $A - B$, $A \cap B$, $B - A$, and $\overline{(A \cup B)}$. The Venn diagram containing three sets divides the universal set into 8 pairwise disjoint regions.

Rather than placing the actual elements in the regions of a Venn diagram, it is sometimes helpful to simply write the cardinality of the set represented by each region. In this way Venn diagrams can be used to evaluate and organize how sets interact, as in the following examples.

Example 3.1.1. In a survey of 300 students, 160 said they play computer games, 80 said they play basketball, and 30 said they play both computer games and basketball. How many students surveyed play neither computer games nor basketball?

Solution. At first glance one might think the answer is $300 - (160 + 80 + 30)$, but then one might realize that the 30 students who said they play both is included in the 160 and in the 80. Let C denote the set of students who play computer games and let B denote the set of students who play basketball. Then the Venn diagram representing the survey results is in Figure 3.2. Since 160 students said they play computer games and 30 said they play both computer games and basketball, 130 students must play computer games but not basketball. So $|C - B| = 130$. Similarly, 50 students play basketball but do not play computer games. The numbers in all 4 regions must add up to 300, so exactly 90 students surveyed play neither computer games nor basketball.

An alternate approach is to assign a variable to each region in the Venn diagram and then solve the resulting system of linear equations. An assign-

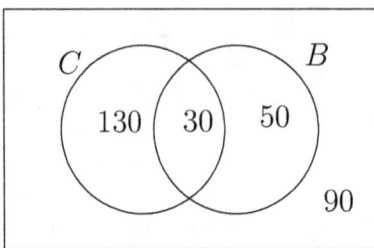

Figure 3.2: Survey results for 300 students

ment as in Figure 3.3 would lead to the equations

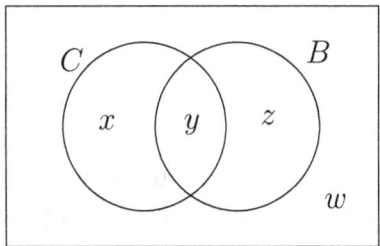

Figure 3.3: Variables representing numbers of students

$$x + y + z + w = 300$$
$$x + y = 160$$
$$y + z = 80$$
$$y = 30$$

This system can easily be solved for the unknowns.   □

Example 3.1.2. The following information was tallied in a survey of 36 students:
- 22 play golf
- 14 play tennis
- 6 play only volleyball

- 9 play golf and tennis
- 6 play tennis and volleyball
- 10 play golf and volleyball
- 2 play all three sports

How many students play at least 2 of the 3 sports? How many play none?

Solution. Since there are 3 sports mentioned, we can start by drawing a Venn diagram containing 3 circles labeled G, T and V (for golf, tennis, and volleyball). Although the information starts by stating that $|G| = 22$, we

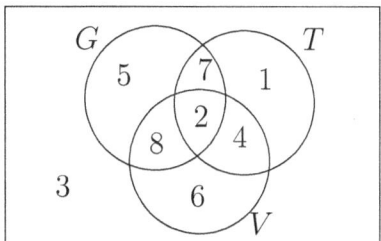

Figure 3.4: Sports played by 36 students

cannot immediately fill in one of the 8 numbers representing the cardinalities of the 8 pairwise disjoint regions; the set G contains 4 of these disjoint regions. We can, however, fill in the region in the center of the diagram, since the last piece of information given tells us that $|G \cap T \cap V| = 2$. Once that is filled in, we can use the fact that $|G \cap V| = 10$ to fill in the cardinality 8 for the region $(G \cap V) - T$. We can continue in this manner, filling in all 8 cardinalities with the given information. Note that because of the word *only* on the third line of the information given, we can immediately fill in the region $V - (G \cup T)$ with a 6. The last region to fill in is the region outside all 3 circles. To get the cardinality of this set, we add up all the other numbers and subtract that sum from 36.

After the Venn diagram has been completed, we can conclude that $4 + 7 + 8 + 2 = 21$ students play at least 2 of the 3 sports, and that 3 play none. $\square$

Exercises

3.1.1. Determine whether each of the following statements is true or false.
(a) $\{1\} \in \{1, 2, 3\}$.

(b) $3 \in \{1, 2, 3, 4\}$.
(c) $\{\} \in \{1, 2, 3\}$.
(d) $\{1, 2, 3\} \in \{1, 2, 3, 4\}$.
(e) $3 \in \{\{1, 2\}, \{3\}\}$.
(f) $0 \in \{0\}$.
(g) $0 \in \{0\} - \{3\}$.

3.1.2. Determine whether each of the following statements is true or false.
(a) $\{a, b, c\} \subseteq \{a, c, b\}$.
(b) $\{\} \subseteq \{1, 2, 3\}$.
(c) $\{\} \subsetneq \{a, b, c\}$.
(d) $\{a, c\} \subsetneq \{a, c\}$.
(e) $\{a, c\} \subsetneq \{a, b\}$.
(f) $\{a, b\} \subseteq \{a, b, c\}$.

3.1.3. Determine whether each of the following statements is true or false.
(a) For all sets A and B, $A \cup B \subseteq B$.
(b) For all sets A and B, $A \cap B \subseteq A$.
(c) For all sets A, $\emptyset \subseteq A$.
(d) For all sets A, $\emptyset \subsetneq A$.
(e) $\{\emptyset\} = \emptyset$.
(f) $0 = |\{0\}|$.
(g) $3 = |\{\emptyset, \{1\}, \{2\}, \{1, 2\}\}|$.

3.1.4. Let $U = \{2, 3, 4, 5, 6, 7, 8\}$, $A = \{2, 4, 6\}$, and $B = \{3, 4, 6, 8\}$. Determine the following.
(a) $\overline{A}$
(b) $\overline{A \cup B}$
(c) $\overline{(A \cap B)}$
(d) $A - B$
(e) $|B - A|$

3.1.5. Use the Venn diagram given in Figure 3.5 to find each of the following sets.
(a) $A \cap (B \cup C)$
(b) $A \cup \overline{B}$
(c) $A - (B \cap C)$
(d) $(A - B) \cap (A - C)$
(e) $\overline{A} \cup \overline{B} \cup \overline{C}$
(f) $(A \cap B) \cup C$

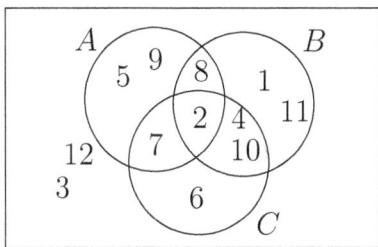

Figure 3.5: Sets for Exercise 3.1.5

3.1.6. Find the power set $\mathcal{P}(S)$, where $S = \{1, 2, 3, 4\}$.

3.1.7. Use induction to prove that for every positive integer n, if $|S| = n$ then $|\mathcal{P}(S)| = 2^n$.

3.1.8. Compute $\mathcal{P}(\emptyset)$ and $\mathcal{P}(\mathcal{P}(\emptyset))$.

3.1.9. Use induction to prove that for every integer $n \geq 2$, if a set S has n elements then S has $n(n-1)/2$ subsets containing exactly two elements.

3.1.10. Pizzazz Pizza offers pizzas with any combination of as many as 6 possible toppings: marshmallows, anchovies, caviar, corn, gummy bears, and coconut. How many different variations of pizzas does Pizzazz Pizza offer? Explain.

3.1.11. Figure 3.6 shows eight disjoint regions in a Venn diagram containing

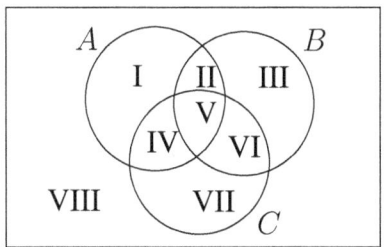

Figure 3.6: Eight regions in a Venn diagram with three sets

three sets A, B and C. Region II can be described using set notation as $(A \cap B) - C$. Give descriptions using set notation for the remaining seven regions.

3.1.12. A survey of 115 randomly selected students gave the following information:
- 15 students are taking only biology
- 50 students are taking English
- 25 students are taking only math
- 15 students are taking only math and biology
- 20 students are taking biology and English
- 25 students are taking only math and English
- 5 students are taking math, English and biology

(a) Construct a Venn diagram and record the appropriate numbers in each region.

(b) How many students surveyed are taking none of the three subjects mentioned?

(c) How many students are taking biology?

(d) How many students are taking English but not math?

3.1.13. At a local club, 36 people were surveyed regarding what sports they play. The following results were compiled:
- 22 play golf
- 14 play tennis
- 20 swim
- 9 play golf and tennis
- 6 play tennis and swim
- 10 play golf and swim
- 2 play golf, play tennis and swim

(a) Of the people surveyed, how many only swim?

(b) How many play exactly two of the sports?

(c) How many play at least two of the sports?

(d) How many play none of the three sports?

(e) How many play golf and tennis, but do not swim?

(f) How many play only one of the three sports?

3.1.14. Destiny Travel surveyed 66 potential customers. The following information was obtained.
- 9 wished to travel only to Hawaii
- 10 wished to travel only to Las Vegas
- 8 wished to travel only to Disney World
- 31 wished to travel to Hawaii
- 10 wished to travel to Hawaii and Las Vegas

- 19 wished to travel to Hawaii and Disney World
- 22 wished to travel to Las Vegas

How many of those surveyed:

(a) did not wish to travel to any of these destinations?

(b) wished to travel to at least two of the destinations?

(c) wished to travel to Las Vegas and Disney World?

3.1.15 (Russell's Paradox). Let T denote the set of all sets which are not elements in themselves; that is, let $T = \{S \mid S \notin S\}$.

(a) Show by contradiction that it cannot be the case that $T \in T$.

(b) Show by contradiction that it cannot be the case that $T \notin T$.

By what you have shown in (a) and (b), it does not make sense to define the set T as we did above. (Defining T the way we did leads to a paradox known as **Russell's Paradox**.) The problem is that we have ignored the context in which a set must be defined; that is, our sets are not all subsets of a given universal set. Given any universal set U, define $T_2 = \{S \mid S \in U \text{ and } S \notin S\}$.

(c) Show that $T_2 \notin T_2$.

(d) Show that $T_2 \notin U$.

By what you have shown in (c) and (d), the use of the universal set in our definition of T_2 avoids the paradox.

3.2 Proof techniques in set theory

In the previous section we saw that two sets A and B are the same if and only if $A \subseteq B$ and $B \subseteq A$. Also, we saw that $A \subseteq B$ means that for all x, if $x \in A$ then $x \in B$. So to prove that $A = B$, we can prove that each is a subset of the other; to prove $A \subseteq B$, we suppose $x \in A$ and somehow deduce from this supposition that $x \in B$. (The term *suppose* is better here than the term *let*, since if A is the empty set, then you cannot let x be in A.) Similarly, to prove $B \subseteq A$, we suppose $x \in B$ and deduce that $x \in A$. Since we are picking an arbitrary point in one of the sets and deducing that this point is also in the other, this method is called the **pick-a-point method** for proving that one set is a subset of another.

The proof of the following theorem will demonstrate this method.

Theorem 3.2.1 (DeMorgan's Laws for Sets). *For every pair of sets A and B,*

(a) $\overline{A \cap B} = \overline{A} \cup \overline{B}$ *and*
(b) $\overline{A \cup B} = \overline{A} \cap \overline{B}$.

Proof. (a) We first show that $\overline{A \cap B} \subseteq \overline{A} \cup \overline{B}$. Suppose $x \in \overline{A \cap B}$. Then it is false that $x \in A \cap B$. That is, it is false that $x \in A$ and $x \in B$. By DeMorgan's Laws for Logic (Theorem 1.2.3), $x \notin A$ or $x \notin B$. Hence, $x \in \overline{A} \cup \overline{B}$.

The proof that $\overline{A} \cup \overline{B} \subseteq \overline{A \cap B}$ is similar and is left as an exercise.

(b) To show $\overline{A \cup B} = \overline{A} \cap \overline{B}$, we show that $\overline{A \cup B} \subseteq \overline{A} \cap \overline{B}$ and $\overline{A} \cap \overline{B} \subseteq \overline{A \cup B}$. To show $\overline{A \cup B} \subseteq \overline{A} \cap \overline{B}$, suppose $x \in \overline{A \cup B}$. Then it is false that $x \in A \cup B$. That is, it is false that $x \in A$ or $x \in B$. By DeMorgan's Laws for Logic, $x \notin A$ and $x \notin B$. Hence, $x \in \overline{A}$ and $x \in \overline{B}$. Thus, $x \in \overline{A} \cap \overline{B}$.

We leave the proof that $\overline{A} \cap \overline{B} \subseteq \overline{A \cup B}$ as an exercise. □

The method of proving that $S \subseteq T$ and $T \subseteq S$ is not the only method for proving that $S = T$. Another way of proving that $S = T$ is to start with the set S, and demonstrate that S is equal to some other set, which in turn is equal to another set, which is in turn equal to T. Usually the sets between S and T in this chain of equalities come from using the **algebra of sets**, which is a collection of properties of unions, intersections, and complements of sets, including DeMorgan's Laws and properties from the following theorem.

Theorem 3.2.2. *Let A, B and C be sets. Then:*
(1) **(Commutative Properties)**
    a. $A \cup B = B \cup A$.
    b. $A \cap B = B \cap A$.
(2) **(Associative Properties)**
    a. $A \cup (B \cup C) = (A \cup B) \cup C$.
    b. $A \cap (B \cap C) = (A \cap B) \cap C$.
(3) **(Distributive Properties)**
    a. $A \cup (B \cap C) = (A \cup B) \cap (A \cup C)$.
    b. $A \cap (B \cup C) = (A \cap B) \cup (A \cap C)$.
    c. $(A \cup B) \cap C = (A \cap C) \cup (B \cap C)$.
    d. $(A \cap B) \cup C = (A \cup C) \cap (B \cup C)$.
(4) $A \cup \emptyset = A$.
(5) $A \cup U = U$.
(6) $A \cup \overline{A} = U$.
(7) $A \cap \emptyset = \emptyset$.
(8) $A \cap U = A$.

(9) $A \cap \overline{A} = \emptyset$.
(10) $A - B = A \cap \overline{B}$.
(11) $\overline{\overline{A}} = A$.

Proof (of (1)a). Since this is the first equality in the algebra of sets, we cannot yet use the algebra of sets to prove (1)a. So we prove that $A \cup B \subseteq B \cup A$ and $B \cup A \subseteq A \cup B$. To show $A \cup B \subseteq B \cup A$, suppose $x \in A \cup B$. Then $x \in A$ or $x \in B$. Thus, $x \in B$ or $x \in A$. It follows that $x \in B \cup A$. Hence, $A \cup B \subseteq B \cup A$. The proof that $B \cup A \subseteq A \cup B$ is similar. **q.e.d.**

Proof (of (3)a). Suppose $x \in A \cup (B \cap C)$. Then $x \in A$ or $x \in (B \cap C)$. So there are two cases to consider. Suppose for the first case that $x \in A$. Then $x \in A \cup B$ and $x \in A \cup C$. Hence, $x \in (A \cup B) \cap (A \cup C)$. Suppose for the second case that $x \in (B \cap C)$. Then $x \in B$ and $x \in C$. Since $x \in B$, $x \in A \cup B$. Since $x \in C$, $x \in A \cup C$. Hence, $x \in (A \cup B) \cap (A \cup C)$. So in both cases, $x \in (A \cup B) \cap (A \cup C)$ and this proves that $A \cup (B \cap C) \subseteq (A \cup B) \cap (A \cup C)$.

  Suppose $x \in (A \cup B) \cap (A \cup C)$. Then $x \in (A \cup B)$ and $x \in (A \cup C)$. There are two possible cases for x: either $x \in A$ or $x \notin A$. In the first case, if $x \in A$ then $x \in A \cup (B \cap C)$. In the second case, if $x \notin A$, then since $x \in (A \cup B)$, it follows that $x \in B$. Similarly, if $x \notin A$, then since $x \in (A \cup C)$, it follows that $x \in C$. Hence, in the case $x \notin A$, x must be in $B \cap C$, so $x \in A \cup (B \cap C)$. **q.e.d.**

  Once we have proven Properties (1)a and (3)a, we can use them in the proofs of other properties. The following proof is our first example of a proof which uses the algebra of sets.

Proof (of (3)d). By the commutative property for unions (Property (1)a), $(A \cap B) \cup C = C \cup (A \cap B)$. By Property (3)a, $C \cup (A \cap B) = (C \cup A) \cap (C \cup B)$. By the commutative property for unions, $C \cup A = A \cup C$ and $C \cup B = B \cup C$. Hence, $(C \cup A) \cap (C \cup B) = (A \cup C) \cap (B \cup C)$. Therefore, $(A \cap B) \cup C = (A \cup C) \cap (B \cup C)$. **q.e.d.**

  When attempting to prove or disprove a possible equality of sets, it might help to draw a Venn diagram of each set and compare the diagrams. If they look the same, then we should attempt to use the pick-a-point method or the algebra of sets to prove that they are indeed the same. If the diagrams do not look the same, then we should attempt to disprove the equality by producing a counterexample. A counterexample would need to consist of very specific sets which do not satisfy the equality.

Example 3.2.3. Prove or disprove: for all sets A and B, $B - (B - A) = A$.

Solution. It looks like the equality might be true because it is true for real numbers: $x - (x - y) = x - x + y = y$. But if we draw Venn diagrams for the sets $B - A$ and $B - (B - A)$, we find that the equality does not necessarily hold; see Figure 3.7. The Venn diagram in Figure 3.7 suggests

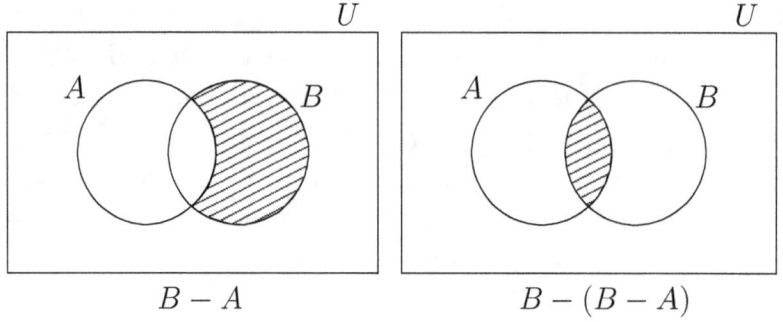

Figure 3.7: Venn diagrams for determining if $B - (B - A) = A$

that if there is an element in A that is not in B, then $B - (B - A)$ and A will be different. For a specific counterexample, let $A = \{1, 2\}$ and let $B = \{2, 3\}$. Then $B - A = \{3\}$, so $B - (B - A) = \{2, 3\} - \{3\} = \{2\}$. Hence, $B - (B - A) \neq A$.   □

Example 3.2.4. Prove that for all sets A, B, and C,

$$\overline{A \cup (B \cap C)} = (\overline{C} \cup \overline{B}) \cap \overline{A}.$$

Proof. We use the algebra of sets. First, by DeMorgan's Law for the complement of a union (3.2.1 (b)),

$$\overline{A \cup (B \cap C)} = \overline{A} \cap \overline{B \cap C}.$$

Next, by DeMorgan's Law for the complement of an intersection (3.2.1 (a)),

$$\overline{A} \cap \overline{B \cap C} = \overline{A} \cap (\overline{B} \cup \overline{C}).$$

By the commutative property for intersections (3.2.2 (1)b),

$$\overline{A} \cap (\overline{B} \cup \overline{C}) = (\overline{B} \cup \overline{C}) \cap \overline{A}.$$

By the commutative property for unions (3.2.2 (1)a),

$$(\overline{B \cup C}) \cap \overline{A} = (\overline{C} \cup \overline{B}) \cap \overline{A}.$$

Hence, by following the string of equalities, we get

$$\overline{A \cup (B \cap C)} = (\overline{C} \cup \overline{B}) \cap \overline{A}.$$

q.e.d.

Exercises

3.2.1. (a) Finish the proof of Theorem 3.2.1 (a).
(b) Finish the proof of Theorem 3.2.1 (b).

3.2.2. (a) Prove Property (1)b of Theorem 3.2.2.
(b) Prove Property (3)b of the same theorem.
(c) Use (a) and (b) to prove Property (3)c.

3.2.3. (a) Use the pick-a-point method to prove $(A \cup B) \cap (A \cup \overline{B}) = A$.
(b) Use the algebra of sets to prove $(A \cup B) \cap (A \cup \overline{B}) = A$.

3.2.4. Prove that if $\overline{A} \subseteq \overline{B}$, then $B \subseteq A$.

3.2.5. Prove that for all sets A and B, $B - (B - A) = A \cap B$.

3.2.6. Prove that for all sets A, B, and C, $(A - B) - C = (A - C) - (B - C)$.

3.2.7. Prove or disprove that for all sets A, B and C,

$$A \cap (B \cup C) = (A \cap B) \cup C.$$

3.2.8. Prove: if $A \subseteq B$ and $C \subseteq D$, then $A \cap C \subseteq B \cap D$.

3.2.9. Prove or disprove: if $A \cap B = A \cap C$, then $B = C$.

3.2.10. Prove or disprove: if $A \cap B = A \cup B$, then $A = B$.

3.2.11. Recall that $\mathcal{P}(S)$ denotes the power set of S. Prove or disprove: $\mathcal{P}(A \cup B) = \mathcal{P}(A) \cup \mathcal{P}(B)$.

3.2.12. Prove or disprove: $\mathcal{P}(A \cap B) = \mathcal{P}(A) \cap \mathcal{P}(B)$.

3.2.13. Prove the following for all sets A and B:
(a) $A \subseteq B$ if and only if $\mathcal{P}(A) \subseteq \mathcal{P}(B)$; and
(b) $A \subsetneq B$ if and only if $\mathcal{P}(A) \subsetneq \mathcal{P}(B)$.

Definition 3.2.5. The **symmetric difference** of two sets A and B is denoted by $A\Delta B$; it is defined to be

$$A\Delta B = (A \cup B) - (A \cap B).$$

3.2.14. Prove that $A\Delta B = (A - B) \cup (B - A)$.

3.2.15. Prove that sets are commutative under symmetric difference:

$$A\Delta B = B\Delta A.$$

3.2.16. Prove that sets are associative under symmetric difference:

$$A\Delta(B\Delta C) = (A\Delta B)\Delta C.$$

3.2.17. One of the following distributive properties is valid and the other is not. Prove that the one which is valid is correct and find a particular counterexample to show that the other one is invalid:

$$A\cup(B\Delta C) \overset{?}{=} (A \cup B)\Delta(A \cup C)$$

$$A\cap(B\Delta C) \overset{?}{=} (A \cap B)\Delta(A \cap C)$$

3.2.18. What is $A\Delta\emptyset$? What is $A\Delta A$?

3.3   Cartesian products and relations

You have probably seen the notation (x, y) used to denote a point in the plane. Since the points (x, y) and (y, x) are usually at different locations, the order in which we write the x and y makes a difference. The expression (x, y) is called an **ordered pair**. The **first coordinate** of (x, y) is x, and the **second coordinate** is y. Two ordered pairs (x, y) and (a, b) are considered equal if and only if $x = a$ and $y = b$.

Definition 3.3.1. Given two sets A and B, the **Cartesian product**, denoted by $A \times B$, is the set of all ordered pairs with first coordinate in A and second coordinate in B; that is,

$$A \times B = \{(a, b) : a \in A \text{ and } b \in B\}.$$

For example, the Cartesian product $\mathbb{R} \times \mathbb{R}$ consists of all ordered pairs of real numbers; this is the plane, which is sometimes denoted by $\mathbb{R}^2$.

Example 3.3.2. Let $A = \{1, 2, 3\}$ and $B = \{\alpha, \gamma\}$. Then

$$A \times B = \{(1, \alpha), (1, \gamma), (2, \alpha), (2, \gamma), (3, \alpha), (3, \gamma)\}.$$

A graphical representation of $A \times B$ is shown in Figure 3.8. $\square$

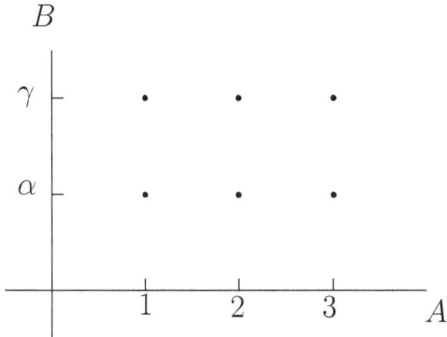

Figure 3.8: A graphical representation of $A \times B$ from Example 3.3.2. The elements of A are listed on the horizontal axis and the elements of B are listed on the vertical axis.

You can see from Example 3.3.2 that if $|A| = n$ and $|B| = m$, then $|A \times B| = nm$.

For a given integer $n > 2$, an *n-tuple* is an ordered list of n objects. Parentheses are used to enclose the list. For example, (x, y, z) is a 3-tuple and $(1, 2, 5, 8)$ is a 4-tuple.

Definition 3.3.3. Given n sets A_1, A_2, ..., A_n, the **Cartesian product**

$$\prod_{j=1}^{n} A_j = A_1 \times A_2 \times \cdots \times A_n$$

is the set of all n-tuples whose j^{th} coordinates are in A_j:

$$\prod_{j=1}^{n} A_j = \{(a_1, a_2, \ldots, a_n) : a_j \in A_j \ \forall j\}.$$

Example 3.3.4. Let $A = \{1, 2\}$, $B = \{a, b, c\}$, and $C = \{\alpha, \beta\}$. Then $A \times B \times C$ consists of twelve 3-tuples:

$$A \times B \times C = \{(1, a, \alpha), (1, a, \beta), (1, b, \alpha), \dots, (2, c, \beta)\}.$$

$\square$

The notion of the Cartesian product of n sets can be extended to the Cartesian product of infinitely many sets. Given a collection of sets $\{A_j : j \in \mathbb{N}\}$, the Cartesian product

$$\prod_{j=1}^{\infty} A_j$$

is the set of all infinite, ordered lists $(a_1, a_2, \dots)$, where $a_j \in A_j$ for all j. An element of this Cartesian product is called an **infinite sequence**. There is much to study about infinite sequences, as you have probably seen in calculus; but for now we return to Cartesian products of just two sets.

Definition 3.3.5. Given a set A and a set B, a **relation from** A **to** B is a subset of $A \times B$. That is, a relation from A to B is a set of ordered pairs where the first coordinates are all in A and the second coordinates are all in B. Given a relation R from A to B, the **domain** of R is the set of all first coordinates of ordered pairs in R; that is,

$$\text{domain}(R) = \{a \in A : \exists b \in B \text{ with } (a, b) \in R\}.$$

Similarly, the **range** of R is the set of all second coordinates of ordered pairs in R:

$$\text{range}(R) = \{b \in B : \exists a \in A \text{ with } (a, b) \in R\}.$$

Example 3.3.6. Let $A = \{a, g, w, z\}$ and $B = \{2, 4, 7\}$. Then

$$R = \{(a, 2), (a, 7), (w, 2)\},$$

$$S = \{(a, 4), (g, 4)\},$$

and

$$T = \{(a, 2), (a, 4), (a, 7), (w, 7), (z, 2), (z, 7)\}$$

are all relations from A to B. The domain of R is $\{a, w\}$ and the range of R is $\{2, 7\}$. The domain of S is $\{a, g\}$ and the range of S is $\{4\}$. The domain of T is $\{a, w, z\}$ and the range is $\{2, 4, 7\}$. $\square$

If R is a relation from A to B and $(a, b) \in R$, we write $a\,R\,b$ and say a *is related to b.*

Example 3.3.7. Let R denote the relation from $\mathbb{R}$ to $\mathbb{R}$ given by

$$R = \{(x, y) : x \text{ is less than } y\}.$$

Then $1\,R\,5$ since 1 is less than 5, and $\pi\,R\,4.2$ since π is less than 4.2. This relation R is the *less than* relation, and it is denoted by the symbol $<$. Instead of writing $1\,R\,5$, we usually write $1 < 5$. □

Example 3.3.8. Let R denote the relation from $\mathbb{Z}$ to $\mathbb{Z}$ given by $n\,R\,m$ if n and m are **relatively prime**; this means that the greatest common divisor of n and m is the integer 1. Then 2 is related to 3 and 3 is related to 4. But 5 is not related to 10 because the greatest common divisor of 5 and 10 is not 1. □

Definition 3.3.9. Given a set A, a **relation on** A is a relation from A to itself.

Definition 3.3.10. Given a set A and a relation R on A, we say:
 R is **reflexive** if $\forall a \in A$, $a\,R\,a$.
 R is **symmetric** if $\forall a, b \in A$, if $a\,R\,b$ then $b\,R\,a$.
 R is **antisymmetric** if $\forall a, b \in A$, if $a\,R\,b$ and $b\,R\,a$, then $a = b$.
 R is **transitive** if $\forall a, b, c \in A$, if $a\,R\,b$ and $b\,R\,c$, then $a\,R\,c$.

Working through some examples is a great way to gain a real understanding of the above definitions.

Example 3.3.11. Let $A = \{1, 2, 3, 4\}$. Then

$$R = \{(1, 2), (2, 1), (2, 3), (4, 4)\}$$

is a relation on A, but R is not reflexive because, for a counterexample, $(1, 1) \notin R$. In order for a relation on A to be reflexive, it must contain the ordered pairs $(1, 1)$, $(2, 2)$, $(3, 3)$ and $(4, 4)$. The relation R is not symmetric because, for a counterexample, $(2, 3) \in R$ but $(3, 2) \notin R$. The relation R is not antisymmetric because $(1, 2) \in R$ and $(2, 1) \in R$, but $1 \neq 2$. Also, R is not transitive because $1\,R\,2$ and $2\,R\,3$ but $(1, 3) \notin R$.

This example shows that *antisymmetric* does not mean the same thing as *not symmetric*. □

Example 3.3.12. The *less than* relation ($<$) on $\mathbb{R}$ given in Example 3.3.7 is not reflexive because numbers are not less than themselves. It is not symmetric because if $a < b$ then b is not less than a. It is antisymmetric because $\forall a, b$, if $a < b$ then b cannot be less than a, so the hypothesis in the definition of *antisymmetric* is never satisfied. It is transitive because if $a < b$ and $b < c$ then $a < c$.

On the other hand, the relation $\leq$ on $\mathbb{R}$ is reflexive since every real number a is less than or equal to itself. The relation $\leq$ is not symmetric; for example, $1 \leq 4$ but 4 is not less than or equal to 1. The relation $\leq$ is antisymmetric, for if $a \leq b$ and $b \leq a$, then $a = b$. The relation $\leq$ is transitive.   □

Example 3.3.13. Let R denote the relation on $\mathbb{R}$ given by

$$R = \{(x, y) \in \mathbb{R} \times \mathbb{R} : y = 3x + 1\}.$$

Then R is not reflexive because $(2, 2) \notin R$. (There are many more examples of ordered pairs (a, a) not in R, but we need only find one example to show that R is not reflexive.) Also, R is not symmetric because $(1, 4) \in R$ but $(4, 1) \notin R$. The relation R is antisymmetric. To see this, suppose $(a, b) \in R$ and $(b, a) \in R$. Then $b = 3a + 1$ and $a = 3b + 1$. Substituting the second equation into the first yields the equality $b = 3(3b + 1) + 1$, which can be solved to yield $b = -\frac{1}{2}$. Since $a = 3b + 1$, $a = -\frac{1}{2}$ also. Hence, $a = b$ and so R is antisymmetric. Finally, the relation R is not transitive because $(1, 4) \in R$ and $(4, 13) \in R$ but $(1, 13) \notin R$.   □

Definition 3.3.14. Let n be a fixed positive integer. Then the integer a is said to be **congruent modulo** n to the integer b if n divides $a - b$. When a is congruent to b modulo n, we write

$$a \equiv b \pmod{n}.$$

Example 3.3.15. Since $23 - 3 = 20$ and $5 \mid 20$, $23 \equiv 3 \pmod 5$. Similarly, $-34 \equiv 22 \pmod 7$, because $-34 - 22 = -56$, and $7 \mid -56$. Recall that $7 \mid -56$ because there is an integer k such that $7 \cdot k = -56$: namely, $k = -8$.   □

Since it takes two integers to be congruent modulo n, this congruence modulo n defines a relation on the set of all integers. The pair (a, b) is in the relation if and only if $a \equiv b \pmod{n}$.

Proposition 3.3.16. *Let n be a fixed positive integer. Then the relation $\equiv$ (mod n) is reflexive, symmetric and transitive.*

Proof. To show that the relation is reflexive, let a be any integer. Then $a - a = 0$, and since $n \cdot 0 = 0$, n must divide $a - a$. Hence, $a \equiv a$ (mod n).

To prove that the relation is symmetric, suppose that $a \equiv b$ (mod n). We must show that $b \equiv a$ (mod n). By definition, if $a \equiv b$ (mod n), then n divides $a - b$. Hence, there exists some integer k such that $n \cdot k = a - b$. But then $b - a = n \cdot (-k)$, and $(-k)$ is an integer also, so n must divide $b - a$. By definition, $b \equiv a$ (mod n).

To prove transitivity, suppose $a \equiv b$ (mod n) and $b \equiv c$ (mod n). We must show that $a \equiv c$ (mod n). Since $a \equiv b$ (mod n), n must divide $a - b$. Hence, there must be some integer k such that $n \cdot k = a - b$. Similarly, since $b \equiv c$ (mod n), n must divide $b - c$, so there must be an integer ℓ such that $n \cdot \ell = b - c$. Well,

$$a - c = (a - b) + (b - c) = n \cdot k + n \cdot \ell = n \cdot (k + \ell).$$

Since $(k + \ell)$ is an integer, n divides $a - c$. By definition, $a \equiv c$ (mod n).

q.e.d.

Exercises

3.3.1. Prove or disprove that for all sets A, B and C,

$$A \times (B \cup C) = (A \times B) \cup (A \times C).$$

3.3.2. Prove or disprove that for all sets A, B, C and D,

$$(A \times B) \cup (C \times D) = (A \cup C) \times (B \cup D).$$

3.3.3. (a) Prove that for all sets X, Y, A and B,

$$(X - A) \times (Y - B) \subseteq (X \times Y) - (A \times B).$$

(b) Give a counterexample to show that it is not true in general that

$$(X \times Y) - (A \times B) \subseteq (X - A) \times (Y - B).$$

3.3.4. Let $A = \{1, 2, 3, 4, 5\}$ and $B = \{2, 4, 6, 7, 9\}$, and let R be the relation from A to B given by

$$R = \{(2, 2), (2, 4), (2, 7), (4, 4), (5, 6)\}.$$

Find the domain and range of R.

3.3.5. Let $A = \{1, 2, 3, 4\}$ and consider the relation R on A given by

$$R = \{(1, 1), (1, 3), (3, 1), (1, 4)\}.$$

Prove or disprove each of the following statements:
(a) R is reflexive.
(b) R is symmetric.
(c) R is antisymmetric.
(d) R is transitive.

3.3.6. Let R denote the relatively prime relation given in Example 3.3.8. Determine if R is reflexive, symmetric, antisymmetric, and transitive, and prove your answers.

3.3.7. Let $A = \{\text{Rock, Paper, Scissors}\}$. What is the relation on A that corresponds to winning combinations of the traditional hand game of *Rock, Paper, Scissors*? (For example, Rock beats Scissors, so (Rock, Scissors) would be a winning combination.) Is the relation reflexive? Symmetric? Antisymmetric? Transitive?

3.3.8. Define a relation R on $\mathbb{R}$ by $x\,R\,y$ if and only if $y = x^2$. Is R reflexive? Why or why not?

3.3.9. Define a relation S on $\mathbb{R}$ by $x\,S\,y$ if and only if $|x + y| = 2$. Is S symmetric? Why or why not?

3.3.10. Define a relation T on $\mathbb{R}$ by $x\,T\,y$ if and only if $x < y + 1$. Is T transitive? Why or why not?

3.3.11. Define a relation U on $\mathbb{R}$ by $x\,U\,y$ if and only if $y = 2x$. Is U antisymmetric? Why or why not?

3.3.12. Prove that if $a \equiv b \pmod{n}$ and $c \equiv d \pmod{n}$, then $a + c \equiv b + d \pmod{n}$.

3.3.13. Prove that if $a \equiv b \pmod{n}$ and $c \equiv d \pmod{n}$, then $ac \equiv bd \pmod{n}$.

3.3.14. Prove that if $a \equiv b \pmod{n}$, then for every positive integer k, $a^k \equiv b^k \pmod{n}$.

3.3.15. Prove that the following claim is false: if $a^k \equiv b^k \pmod{n}$ and $k \equiv \ell \pmod{n}$, then $a^\ell \equiv b^\ell \pmod{n}$.

3.3.16. Prove that the following claim is false: if $a \cdot b \equiv 0 \pmod{n}$, then $a \equiv 0 \pmod{n}$ or $b \equiv 0 \pmod{n}$.

3.4 Equivalence relations and orders

Certain types of relations occur frequently in mathematics. First, when we want to provide a way for elements of a set to be considered the same, or similar, or equivalent in some way, we do so by defining a particular kind of relation called an *equivalence relation* on the set. On the other hand, when we want to provide a way to compare the elements of a set with each other and think of one element as greater, bigger, larger, or better than the other in some way, we do so by defining a relation called an *order* on the set. We study these kinds of relations in this section, beginning with a rigorous definition of an equivalence relation.

Definition 3.4.1. Given a set A, a relation R on A is called an **equivalence relation** on A if R is reflexive, symmetric and transitive.

Equivalence relations are required to be reflexive because in any reasonable use of the term *equivalent*, we would expect that every element should be equivalent to itself. Similarly, symmetry is required because if an element is the same in some way as another, then the second should be the same as the first in that same way. Transitivity is required because if one element is the same in some way as a second element and the second is the same as a third, then the first should be the same as the third.

Example 3.4.2. In geometry, we say that two triangles are **congruent** if they have the same size and shape. (This means that even though the two triangles may not be in the same location, we can perhaps move one triangle, and perhaps rotate it or reflect it, and place it perfectly on top of the other.)

Congruence is an equivalence relation on the set of all triangles in the plane. To see this, you should convince yourself first that congruence is a relation on the set of triangles. Then, note that every triangle is congruent to itself, so that the relation is reflexive. To see that the relation is symmetric, note that if triangle T_1 is congruent to triangle T_2, then T_2 is congruent to T_1. To see that the relation is transitive, note that if triangle T_1 is congruent to T_2 and T_2 is congruent to T_3, then T_1 is congruent to T_3.

Triangle T_1 is said to be **similar** to triangle T_2 if the measure of all three angles in T_2 is the same as the measure of the angles in T_1. Similarity is also an equivalence relation on the set of all triangles. (Why?) □

Proposition 3.4.3. *Modular congruence (the relation $\equiv$ (mod n)) is an equivalence relation on the set of integers.*

Proof. This proposition follows immediately from the definition of equivalence relation and Proposition 3.3.16.   **q.e.d.**

Definition 3.4.4. Given an equivalence relation R on a set A, the **equivalence class** of an element $a \in A$ is the set

$$[a] = \{x \in A : x\,R\,a\}.$$

The element a is called the **representative** of $[a]$.

The equivalence class $[a]$ is the set of all elements in A which are related to a. This includes the representative a itself, since every equivalence relation is reflexive.

Example 3.4.5. Let $A = \{1, 2, 3, 4\}$ and let R be the relation on A given by

$$R = \{(1,1), (2,2), (3,3), (4,4), (2,3), (3,2)\}.$$

Then R is an equivalence relation, since it is reflexive, symmetric and transitive. We can thus compute equivalence classes:

$$
\begin{aligned}
[1] &= \{1\} \\
[2] &= \{2,3\} \\
[3] &= \{2,3\} \\
[4] &= \{4\}
\end{aligned}
$$

Note that the equivalence classes $[2]$ and $[3]$ are the same, even though the representatives 2 and 3 are different.   □

As you can see from Example 3.4.5, the equivalence classes $[x]$ and $[y]$ are either disjoint or the same. In addition, the union of all of the equivalence classes equals the whole set A.

Definition 3.4.6. Given a set A, a collection of subsets of A is called a **partition** of A if every pair of distinct subsets in the collection is disjoint and the union of the sets in the collection equals A.

Hence, the collection of all equivalence classes forms a partition of A. On the other hand, given any partition of a set A, we can define an equivalence relation R on A by $a\,R\,b$ if and only if a and b are elements of the same subset in the partition.

Example 3.4.7. Let $A = \{1, 2, 3, 4, 5, 6\}$ and consider the partition of A given by

$$\{\ \{1, 3\}, \{2\}, \{4, 5\}, \{6\}\ \}.$$

This partition defines the equivalence relation R on A given by

$$R = \{(1, 1), (3, 3), (1, 3), (3, 1), (2, 2), (4, 4), (5, 5), (4, 5), (5, 4), (6, 6)\}.$$

□

Definition 3.4.8. Given an equivalence relation R on a set A, the set of all distinct equivalence classes is called A **modulo** R and denoted by A/R.

Example 3.4.9. If A and R are given as in Example 3.4.7, then

$$A/R = \{[1], [2], [4], [6]\}.$$

□

Example 3.4.10. Consider the relation $\equiv$ (mod 5) on $\mathbb{Z}$. With this equivalence relation, the partition of $\mathbb{Z}$ consists of exactly 5 subsets of $\mathbb{Z}$:

$$
\begin{aligned}
[0] &= \{0, 5, 10, 15, ..., -5, -10, -15, -20, ...\} \\
[1] &= \{1, 6, 11, 16, ..., -4, -9, -14, -19, ...\} \\
[2] &= \{2, 7, 12, 17, ..., -3, -8, -13, -18, ...\} \\
[3] &= \{3, 8, 13, 18, ..., -2, -7, -12, -17, ...\} \\
[4] &= \{4, 9, 14, 19, ..., -1, -6, -11, -16, ...\}
\end{aligned}
$$

Thus, $\mathbb{Z}/\equiv$ is the set $\{[0], [1], [2], [3], [4]\}$. □

Example 3.4.11. Let $A = \mathbb{R} - \{0\}$ and define an equivalence relation R on A by $x\,R\,y$ if $xy > 0$. Show that R is an equivalence relation on A, describe the equivalence classes, and find A/R.

Solution. To show that R is an equivalence relation, we show that R is reflexive, symmetric and transitive. To show R is reflexive, let $x \in A = \mathbb{R} - \{0\}$. Then $xx = x^2 > 0$, so $x\,R\,x$. To show R is symmetric, suppose $x \in A$ and $y \in A$ and $x\,R\,y$. Then $xy > 0$, so $yx > 0$ too. Hence, $y\,R\,x$. To show R is transitive, suppose $x, y, z \in A$ and $x\,R\,y$ and $y\,R\,z$. Then $xy > 0$ and $yz > 0$. Hence, $(xy)(yz) > 0$, and so $(xz)y^2 > 0$. Since $y^2 > 0$ automatically, $xz > 0$ too. Thus, $x\,R\,z$.

The equivalence class of $x = 1$ is the set of all numbers y in A such that $1 \cdot y > 0$; that is, $[1] = \{y : y > 0\}$. Similarly, if $-1 \cdot y > 0$ then $y < 0$, so $[-1] = \{y : y < 0\}$. Since $[1] \cup [-1] = A$, these are the only two equivalence classes, and $A/R = \{[-1], [1]\}$. (Note that $[1] = [x]$ for any $x > 0$, and we would also be correct to state that $A/R = \{[-2], [3]\}$ or $A/R = \{[-\pi], [18.1]\}$.) $\square$

Definition 3.4.12. Given a set A, a relation R on A is called a **partial order** on A if R is reflexive, antisymmetric and transitive.

Example 3.4.13. The relation $\leq$ is a partial order on $\mathbb{R}$, since it is clearly reflexive, antisymmetric and transitive. However, the relation $<$ on $\mathbb{R}$ is not a partial order because this relation is not reflexive. $\square$

Example 3.4.14. Consider the relation on $\mathbb{N}$ given by a is related to b if $a \mid b$. This relation is reflexive, antisymmetric, and transitive, so it is a partial order on $\mathbb{N}$. (See Exercise 3.4.12.) $\square$

Definition 3.4.15. Let R be a partial order on the set A, and let $a \in A$ and $b \in A$. Then we say that a and b are **comparable** if $a\,R\,b$ or $b\,R\,a$. Also, R is a **total order** on A if every pair of elements of A are comparable.

Example 3.4.16. The *divides* relation $\mid$ on $\mathbb{N}$ is not a total order on $\mathbb{N}$ since 2 and 3 are not comparable; that is, 2 does not divide 3 and 3 does not divide 2. However, the *less than or equal to* relation $\leq$ on $\mathbb{R}$ is a total order on $\mathbb{R}$, since for any pair of real numbers x and y, $x \leq y$ or $y \leq x$. $\square$

Exercises

3.4.1. Let $\mathcal{D}$ denote the set of all differentiable functions. Define a relation R on $\mathcal{D}$ by $f\,R\,g$ if and only if $f' = g'$. Prove that R is an equivalence relation on $\mathcal{D}$, and find the equivalence class $[3x]$.

3.4.2. Let $S = \{2, 4, 6, 8\}$ and define a relation R on $\mathcal{P}(S)$ (the power set of S) by

$$A\,R\,B \text{ if and only if } |A| = |B|.$$

Show that R is an equivalence relation on $\mathcal{P}(S)$, and find the equivalence class $[\{2, 6\}]$. Then find $\mathcal{P}(S)/R$.

3.4.3 (Linear Algebra required)**.** If A and B are $n \times n$ matrices, then A is said to be **similar** to B if there exists an $n \times n$ invertible matrix P such that $A = PBP^{-1}$. Prove that similarity is an equivalence relation on the set of all $n \times n$ matrices.

3.4.4. Let $A = \mathbb{Z} \times (\mathbb{Z} - \{0\})$, and define a relation R on A by $(a, b)\,R\,(c, d)$ if and only if $ad = bc$.
(a) Prove that R is an equivalence relation on A.
(b) Relate this to a concept that you have seen before. Why do we require the second coordinates to be nonzero?

3.4.5. Define a relation R on $\mathbb{R}^2$ by $(x, y)\,R\,(a, b)$ if $x^2 + y^2 = a^2 + b^2$. Prove this is an equivalence relation on $\mathbb{R}^2$. Then find the equivalence class $[(0, 2)]$, and interpret this geometrically in the plane.

3.4.6. Define a relation R on $\mathbb{R}^2$ by $(x, y)\,R\,(a, b)$ if $xy = ab$. Prove this is an equivalence relation on $\mathbb{R}^2$. Then find the equivalence classes $[(0, 2)]$ and $[(2, 3)]$, and interpret these geometrically in the plane.

3.4.7. Define a relation $\sim$ on $\mathbb{R}$ by $x \sim y$ if $x - y \in \mathbb{Z}$. Prove that $\sim$ is an equivalence relation on $\mathbb{R}$. Then find the equivalence classes $[0]$ and $[\pi]$. Finally, determine the set $\mathbb{R}/\sim$.

3.4.8. (a) Prove that $48 \equiv 72 \pmod{12}$.
(b) Find the equivalence class $[48]$ for the relation $\equiv \pmod{12}$ on $\mathbb{Z}$.

3.4.9. Consider the relation $\equiv \pmod 7$ on $\mathbb{Z}$. Describe the equivalence classes $[1]$ and $[2]$, and find $\mathbb{Z}/\equiv$.

3.4.10. Suppose n is a positive integer, and consider the relation $\equiv$ (mod n) on $\mathbb{Z}$. Prove that for all integers x, y, z and w, if $[x] = [z]$ and $[y] = [w]$, then $[x + y] = [z + w]$.

3.4.11. Let $A = \{\alpha, \beta, \gamma, \delta\}$ and consider the partition of A given by

$$\{\,\{\alpha, \beta, \gamma\}, \{\delta\}\,\}.$$

Define an equivalence relation R on A by $a\,R\,b$ iff a and b are elements of the same subset in the partition. Give a complete list of the elements of the relation R.

3.4.12. Consider the relation on $\mathbb{N}$ given by a is related to b if a divides b. Prove that this is a partial order on $\mathbb{N}$. That is, prove that this relation is reflexive, antisymmetric, and transitive.

3.4.13. Let S be any set. Prove that $\subseteq$ is a partial order on the power set $\mathcal{P}(S)$. Prove or disprove that $\subseteq$ is a total order on $\mathcal{P}(S)$.

3.4.14. Define a relation R on $\mathbb{R}^2$ by $(x, y)\,R\,(a, b)$ if $x^2 + y^2 \geq a^2 + b^2$. Prove this is not a partial order on $\mathbb{R}^2$.

Summary of proof techniques

This is a good time for a quick review of all of the proof techniques we have covered.

When you are trying to figure out how to prove a statement, you should first identify the type of statement you are dealing with. Once you have identified the type of the statement you can proceed with an appropriate method.

- For a **conditional statement** *if p then q*, there are three appropriate methods:

    - In a **direct proof**, you suppose that the hypothesis p is true and attempt to deduce that the conclusion q is true.

    - In a **proof by contrapositive**, you suppose that the conclusion q of the conditional statement is false and attempt to deduce that the hypothesis p is false.

    - In a **proof by contradiction**, you suppose that the hypothesis p is true and that the conclusion q is false. You then try to deduce a contradiction, which can be any statement that you know must be false.

- If your statement is a **biconditional statement** *p if and only if q*, then most often you should proceed to prove the two conditional statements *if p then q* and *if q then p* independently. Each of these conditional statements can be proven directly, by contrapositive, or by contradiction.

- If your statement is an **equality (=)** or **containment ($\subseteq$) of sets**, there are two appropriate methods:

- When it looks like you might be able to use some previously proven properties of sets, you can use the **algebra of sets** method. In this method you start with the set on one side and use properties like DeMorgan's Laws for Sets or the associative, commutative, or distributive laws to construct a string of equalities or containments that lead to the set on the other side.

- You can use the **pick-a-point** method to prove that one set is a subset of another. In this method, you suppose there is an element in one set and somehow deduce that the point must also be in the other set. To prove that two sets are equal, you can show that each set is a subset of the other.

- For an **existentially quantified predicate**, there are two general methods.

   - In a **constructive proof**, you explicitly construct the item you want to prove exists. Then you simply check that it satisfies all the requirements of the predicate.

   - In a **non-constructive proof**, you usually use previously proven theorems to prove the existence of the item without actually constructing the item which works. This might require advanced tools like the Intermediate Value Theorem.

- For a predicate which has been **existentially quantified with uniqueness** ($\exists!$), first prove existence using a constructive or a nonconstructive proof. Then prove uniqueness by supposing there are two values x, y which make the predicate true, and somehow deducing that x and y must be equal.

- For a **universally quantified predicate**, there are not many options.

   - If the universe of discourse for the universally quantified predicate is a set of integers greater than some value, then you should consider using **mathematical induction**.

   - If the universe of discourse is finite, you can certainly take each case one at a time.

   - In many other circumstances, the best you can hope for is to begin by saying, *Let x be given* [*in the universe of discourse*]. For example, if your universe of discourse is the set of all real numbers, you can begin by saying, *Let x be any real number.*

- If your statement is a **simple proposition**, you might consider the following options.

  - Sometimes it seems natural to ask, *What would go wrong if my statement were not true?* In these cases (and perhaps some others), you might try to prove your statement **by contradiction**.
  - For many simple propostions, the proof will use the **definitions** of the terms involved or some previously proven theorems. For example, to prove that a relation is an equivalence relation, we would naturally use the definition of an equivalence relation.

- If you encounter a **disjunction in a supposition**, or if you can prove your statement more easily under certain circumstances, you should consider a **proof by cases**. For example, if you think you can prove your statement in the case where n is even, then you can go ahead and prove your result for that case, and consider the other cases afterwards.

- If you need to **prove a disjunction** $p \lor q$, use the logical equivalence $p \lor q \Leftrightarrow \sim p \rightarrow q$ (Proposition 1.2.4) and prove the conditional proposition $\sim p \rightarrow q$ using a direct proof, a proof by contrapositive or a proof by contradiction.

Note. Some proofs can be organized into different levels, and sometimes you will have to use different techniques in one proof. You might begin a proof of a conditional statement by using a direct proof; and if the conclusion of the conditional statement is an equality of sets, you might then proceed to use the pick-a-point method. Once you get started in your pick-a-point method, you might find that a proof by cases works best. And in one of the cases you might find that a proof by contradiction works. In the other case you might prefer to use another method.

Example. Let's see how we might go about proving the following statement: *For all real numbers y, if y > 6 then* $\{x : x > y\} \cup \{x : x < -y\} \subseteq \{x : x^2 > 36\}$. The big picture is that the statement is a universally quantified predicate, and our domain of discourse is the set of all real numbers. Thus, we might start with the statement, *Let y be any real number.* Then we're ready to tackle the next part.

We can now concentrate on the statement, *if $y > 6$ then $\{x : x > y\} \cup \{x : x < -y\} \subseteq \{x : x^2 > 36\}$*. Since this is a conditional statement, we might choose to proceed with a direct proof. We can begin the next part like this: *Suppose $y > 6$*. Next, we can tackle the conclusion of the conditional statement.

We would now like to prove the statement, $\{x : x > y\} \cup \{x : x < -y\} \subseteq \{x : x^2 > 36\}$. Since this is a containment of sets, the pick-a-point method is appropriate. Thus, our next sentence can be, *Suppose $w \in \{x : x > y\} \cup \{x : x < -y\}$*. At this stage, we explain what it means for our point w to be in the union: *Then $w \in \{x : x > y\}$ or $w \in \{x : x < -y\}$*. Since this is a disjunction, we proceed to use a proof by cases. We prove the conclusion in the case $w \in \{x : x > y\}$ and then in the other case $w \in \{x : x < -y\}$. We have compiled the complete proof below.

Proof. Let y be any real number. Suppose $y > 6$. Suppose

$$w \in \{x : x > y\} \cup \{x : x < -y\}.$$

Then $w \in \{x : x > y\}$ or $w \in \{x : x < -y\}$.

Case 1: Suppose $w \in \{x : x > y\}$. Then $w > y$. Since $y > 6$ we get $w > 6$. Since $w > 6$, $w^2 > 36$. Hence, $w \in \{x : x^2 > 36\}$.

Case 2: Suppose $w \in \{x : x < -y\}$. Then $w < -y$. Since $y > 6$, $-y < -6$ and $w < -6$. Since $w < -6$, $w^2 > 36$ and $w \in \{x : x^2 > 36\}$.   **q.e.d.**

Note how we have used a variety of proof techniques all in one proof. Using a variety of techniques is common in advanced proofs.   □

Chapter 4

Functions

Fundamental progress has to do with the reinterpretation of basic ideas.

-Alfred North Whitehead, philosopher and mathematician (1861-1947)

4.1 Functions, injectivity and surjectivity

The concept of a function is so central to mathematics that functions are studied in virtually every branch of math. Even so, the function concept is not a simple one and it can take a long time and a lot of study to completely understand.

In high school or college algebra you may have seen the term *function* defined as a rule or correspondence from one set called the *domain*, to another set called the *range* such that each member of the domain corresponds to exactly one member of the range. The term *rule* or *correspondence* may or may not have been precisely defined, but in this situation a *rule* or *correspondence* from one set A to another set B is just a relation from A to B:

Definition 4.1.1. A **function** f from the set A to the set B is a relation from A to B such that the domain is all of A and such that if $(a, b) \in f$ and $(a, c) \in f$, then $b = c$. When f is a function from A to B, then B is the **codomain** of f, and we write $f : A \to B$. If $(a, b) \in f$, we usually denote this by writing $f(a) = b$.

Example 4.1.2. Let $A = \{1, 2, 3\}$ and $B = \{\alpha, \beta\}$. Then $f = \{(1, \alpha), (2, \beta)\}$ is a relation from A to B, but f is not a function from A to B since the domain of f is not all of A. Also, $g = \{(1, \alpha), (2, \beta), (2, \alpha), (3, \beta)\}$ is a relation from A to B but g is not a function from A to B because $(2, \beta) \in g$ and $(2, \alpha) \in g$ but $\alpha \neq \beta$. However, $h = \{(1, \alpha), (2, \beta), (3, \alpha)\}$ is a function from A to B. $\square$

Example 4.1.3. Let $A = \{a, b, c, d, e\}$ and let $B = \{1, 2, 3, 4\}$. Then $f = \{(a, 3), (b, 2), (c, 3), (d, 2), (e, 3)\}$ is a function from A to B. The domain of f is A, the codomain of f is B, and the range of f is $\{2, 3\}$. Figure 4.1 shows

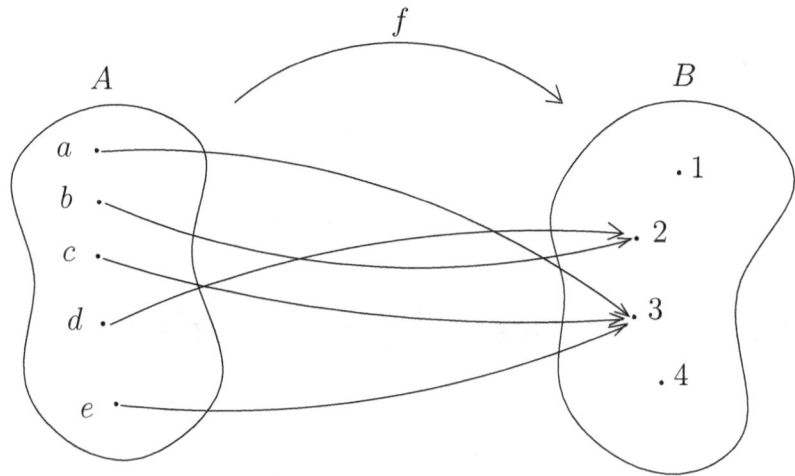

Figure 4.1: An arrow diagram for the function in Example 4.1.3.

a drawing of the function f. An arrow is used to represent each ordered pair in f. Such a drawing is called an **arrow diagram**.   Arrow diagrams can be used to represent not only functions but also more general relations. A relation from A to B is a function if and only if in its arrow diagram, (1) there is an arrow beginning at each point of A; and (2) there are not two arrows beginning at the same point in A but ending at different points in B. $\square$

Example 4.1.4. Let $A = \{5, 12\}$ and let $B = \{\alpha, \beta, \gamma\}$. Then there are exactly 9 distinct functions from A to B:

f	$f(5)$	$f(12)$
f_1	α	α
f_2	α	β
f_3	α	γ
f_4	β	α
f_5	β	β
f_6	β	γ
f_7	γ	α
f_8	γ	β
f_9	γ	γ

In the chart above, f_3 is the function $f_3 = \{(5, \alpha), (12, \gamma)\}$. □

If A and B are subsets of $\mathbb{R}$ and $f : A \to B$, then the requirement that *if $(a, b) \in f$ and $(a, c) \in f$ then $b = c$* means that the graph of f passes the **Vertical Line Test**, which states that the graph of an equation is the graph of a function if and only if no vertical line intersects the graph more than once.

Example 4.1.5. Consider the graph of the equation $x^2 + y^2 = 1$, the unit circle. Since the vertical line $x = 0$ passes through this circle in more than one point, the graph of the equation $x^2 + y^2 = 1$ is not the graph of a function. That is, the graph passes through the points $(0, -1)$ and $(0, 1)$, which have the same first coordinates but different second coordinates. But for any function, the second coordinate is completely determined by the first coordinate.

When we solve the equation $x^2 + y^2 = 1$ for y, we get two solutions: $y = \sqrt{1 - x^2}$ and $y = -\sqrt{1 - x^2}$. The graph of each of these solutions does pass the Vertical Line Test; the graph of $y = \sqrt{1 - x^2}$ is the upper half of the unit circle, and the graph of $y = -\sqrt{1 - x^2}$ is the lower half. Independently, each of these graphs is the graph of a function. □

In algebra and calculus when we write a formula for a function like

$$f(x) = \frac{x + 2}{x - 1} + 3,$$

it is assumed that the codomain is $\mathbb{R}$ and that the domain is the subset of $\mathbb{R}$ consisting of those real numbers that yield real outputs when they are plugged into the formula. The domain of the function $f(x)$ above is $\{x \in \mathbb{R} : x \neq 1\}$. The range of f is $\{y \in \mathbb{R} : y \neq 4\}$; this can be seen from the graph of $y = f(x)$.

Definition 4.1.6. Let f be a function from A to B. Then f is **injective** or **one-to-one** if the following conditional proposition is satisfied:

$$\forall a, b \in A \ (f(a) = f(b)) \rightarrow (a = b).$$

An injective function is called an **injection**.

To prove that a function $f : A \rightarrow B$ is one-to-one, we can give a direct proof of the conditional statement above: suppose a and b are in A and $f(a) = f(b)$, and try to deduce that $a = b$ using some knowledge of the function f. Or, we can try a proof by contrapositive, where we suppose $a \neq b$, and try to deduce that $f(a) \neq f(b)$.

Example 4.1.7. Let $f : \mathbb{R} \rightarrow \mathbb{R}$ be defined by $f(x) = 1 - 2x$. Prove that f is one-to-one.

Proof. Let a and b be real numbers, and suppose $f(a) = f(b)$. Then $1 - 2a = 1 - 2b$. Subtracting 1 from both sides of the equation yields the equation $-2a = -2b$. Dividing by -2 gives us $a = b$. Hence, if $f(a) = f(b)$, then $a = b$; so f is one-to-one.   **q.e.d.**

□

Example 4.1.8. Let $f : \mathbb{R} \rightarrow \mathbb{R}^2$ be given by the formula $f(x) = (x^2, x^3)$. Prove that f is one-to-one.

Proof. Suppose $a, b \in \mathbb{R}$ and $f(a) = f(b)$. Then $(a^2, a^3) = (b^2, b^3)$. Therefore, $a^2 = b^2$ and $a^3 = b^3$. By taking the cube root of both sides of the equation $a^3 = b^3$, we get $a = b$.   **q.e.d.**

Note that in the above proof we did not use the fact that $a^2 = b^2$. Indeed, we could not have concluded that f was one-to-one by using only the equation $a^2 = b^2$, for this equation implies that $a = \pm b$, not that $a = b$.   □

A function f is one-to-one if the conditional proposition

if $(a, c) \in f$ and $(b, c) \in f$ then $a = b$

is satisfied. For a function $f : A \to \mathbb{R}$ where $A \subseteq \mathbb{R}$, this means that the graph passes the **Horizontal Line Test**: a function f from a subset of $\mathbb{R}$ to $\mathbb{R}$ is one-to-one if and only if no horizontal line intersects the graph of f more than once.

In the arrow diagram for a function, the function is one-to-one if and only if there are not two different arrows that point to the same element in the codomain. Hence, the function pictured in Figure 4.1 is not one-to-one because there are two different arrows pointing to the number 2. Likewise, there also exist two different arrows pointing to the number 3.

To prove that a function $f : A \to B$ is not one-to-one, we can prove the negation of the proposition in Definition 4.1.6:

$$\exists a, b \in A \ (f(a) = f(b) \text{ and } a \neq b).$$

To prove this existentially quantified predicate, it suffices to give one particular example of elements a and b in A for which $a \neq b$ and $f(a) = f(b)$.

Example 4.1.9. Prove that the function $f : \{5, 6, 7\} \to \{1, 2\}$ given by $f = \{(5, 1), (6, 2), (7, 2)\}$ is not one-to-one.

Proof. From the definition of the function f we see that $f(6) = 2$ and $f(7) = 2$. Since $f(6) = f(7)$ but $6 \neq 7$, f is not one-to-one.   **q.e.d.**

□

Example 4.1.10. Let P denote the set of all polynomial functions with real coefficients. Let D be the function from P to P such that for any $f(x) \in P$, $D(f(x)) = f'(x)$ (the derivative of $f(x)$). Then the function D is not one-to-one. To prove this, it suffices to find two distinct polynomials which have the same derivative. For example, $D(2x^3 + 3x - 2) = 6x^2 + 3 = D(2x^3 + 3x + 7)$, yet $2x^3 + 3x - 2 \neq 2x^3 + 3x + 7$.   □

Definition 4.1.11. Let $f : A \to B$. Then f is **surjective** or **onto** if the range of f is equal to the codomain B. A surjective function is called a **surjection**.

Since the range of a function $f : A \to B$ is always a subset of the codomain, to prove that f is onto it is enough to prove that the codomain is a subset of the range. The pick-a-point method can be used for this: suppose b is in the codomain B, and deduce (using some knowledge of the function f) that b must also be in the range of f.

Example 4.1.12. Let $f : \mathbb{R} \to \mathbb{R}$ be defined by $f(x) = 1 - 2x$. Prove that f is onto.

Solution. Using the pick-a-point method, we will pick a point y in the codomain $\mathbb{R}$, and we will try to find an x in the domain $\mathbb{R}$ such that $f(x) = y$. Solving the equation $1 - 2x = y$ for x yields $x = (1 - y)/2$, so this will be the value of x which works in the proof.

Proof. Suppose $y \in \mathbb{R}$. Let $x = (1 - y)/2$. Then x is a real number, so x is in the domain of f. Furthermore, $f(x) = 1 - 2x = 1 - 2(\,(1 - y)/2\,) = y$. Thus, the codomain of f is a subset of the range of f, so f must be onto.   **q.e.d.**

$\square$

Example 4.1.13. Let $f : \mathbb{R}^2 \to \mathbb{R}$ be given by $f(x, y) = \frac{1}{2}x + y^3$. Prove that f is onto.

Proof. Suppose $b \in \mathbb{R}$. Let $(x, y) = (2b, 0)$. Then $(x, y) \in \mathbb{R}^2$, and $f(x, y) = f(2b, 0) = \frac{1}{2}(2b) + 0^3 = b$.   **q.e.d.**

$\square$

In the arrow diagram for a function, the function is onto if and only if each element of the codomain has an arrow pointing to it. The function in Figure 4.1 is not onto because there is no arrow pointing to the number 1. Likewise, there is no arrow pointing to the number 4.

To prove that a function $f : A \to B$ is not onto, we can try to find an element of the codomain B that is not in the range of f.

Example 4.1.14. Let $f : \mathbb{R} \to \mathbb{R}$ be given by the formula $f(x) = x^2$. Prove that f is not onto.

Proof. Let $y = -1$. Then since y is a real number, y is in the codomain of f. But since $f(x) \geq 0$ for all $x \in \mathbb{R}$, there is no $x \in \mathbb{R}$ such that $f(x) = -1$. Hence, $y = -1$ is in the codomain of f but not in the range of f. Therefore, f is not onto.   **q.e.d.**

Example 4.1.15. Let $f : \mathbb{R} \to \mathbb{R}^2$ be given by $f(x) = (x, x^3)$. Prove that f is not surjective.

Proof. Since $2 \in \mathbb{R}$ and $3 \in \mathbb{R}$, the ordered pair $(2, 3) \in \mathbb{R}^2$. However, $(2, 3)$ is not in the range of f, for if $f(x) = (2, 3)$, then $(x, x^3) = (2, 3)$ and $x = 2$ and $x^3 = 3$. But if $x = 2$ then $x^3 = 8$, not 3. By contradiction, there can be no x such that $f(x) = (2, 3)$.   **q.e.d.**

Of course the point $(2, 3)$ is nothing special, except that it is not in the range of f; many other points would work as well. ☐

Definition 4.1.16. Let $f : A \to B$. Then f is **bijective** if f is injective and surjective. Such a function is called a **bijection**.

Example 4.1.17. Let $f : \mathbb{R} \to \mathbb{R}$ be given by $f(x) = 1 - 2x$. We saw in Example 4.1.7 that f is injective, and in Example 4.1.12 that f is surjective. Hence, f is a bijection from $\mathbb{R}$ to $\mathbb{R}$. ☐

Definition 4.1.18. Let $f : A \to B$ and $g : B \to C$. Then the **composition** $g \circ f$ is defined to be the function from A to C given by

$$g \circ f = \{(a, c) : \exists b \in B \text{ with } (a, b) \in f \text{ and } (b, c) \in g\}.$$

Example 4.1.19. Let $A = \{1, 2, 3\}$, $B = \{u, v, w, y\}$ and $C = \{3, 4, 7, 9\}$. Let $f : A \to B$ and $g : B \to C$ be given by

$$f = \{(1, u), (2, w), (3, v)\}$$

and

$$g = \{(u, 9), (v, 4), (w, 4), (y, 3)\}.$$

Then we can visualize the function $g \circ f$ from A to C by first applying f and then applying g. That is, $(g \circ f)(a) = g(f(a))$. For example, $(g \circ f)(1) = g(f(1)) = g(u) = 9$. We can find $(g \circ f)(2)$ and $(g \circ f)(3)$ also to get the function $g \circ f$ as a list of ordered pairs:

$$g \circ f = \{(1, 9), (2, 4), (3, 4)\}.$$

This list of ordered pairs is apparent from the combined arrow diagrams for f and g. The ordered pairs in $g \circ f$ can be found by following the arrows in Figure 4.2 from A all the way to C. Note that 3 is in the range of g but 3 is not in the range of $g \circ f$. ☐

The next couple of theorems say that the composition of surjective functions is surjective and that the composition of injective functions is injective. These are important theorems and you will see them again and again in your future studies in mathematics, in abstract algebra, topology, and the like.

Theorem 4.1.20. *Suppose $f : A \to B$ is onto and $g : B \to C$ is onto. Then the composition $g \circ f : A \to C$ is onto.*

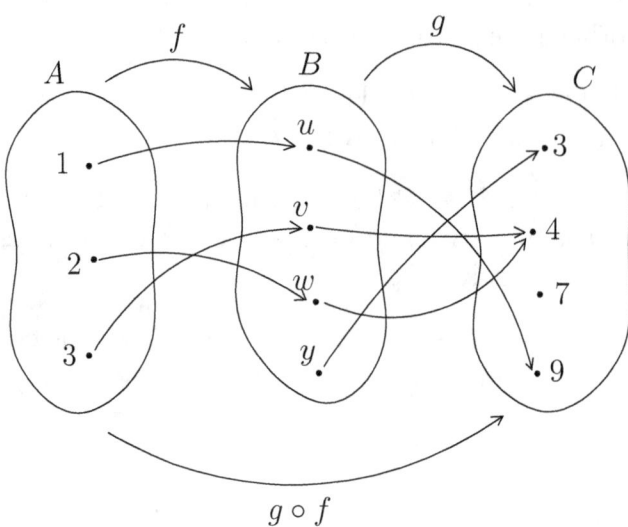

Figure 4.2: The composition of functions from Example 4.1.19.

Proof. Suppose $f : A \to B$ is onto and $g : B \to C$ is onto. Suppose $c \in C$. We must show that there is some $a \in A$ such that $(g \circ f)(a) = c$. Since $g : B \to C$ is onto, there must be some $b \in B$ such that $g(b) = c$. Since $f : A \to B$ is onto, there must be some $a \in A$ such that $f(a) = b$. Hence, $(g \circ f)(a) = g(f(a)) = g(b) = c$. Thus, $g \circ f$ is onto.   **q.e.d.**

Theorem 4.1.21. *Suppose $f : A \to B$ is one-to-one and $g : B \to C$ is one-to-one. Then $g \circ f : A \to C$ is one-to-one.*

Proof. Suppose $f : A \to B$ is one-to-one and $g : B \to C$ is one-to-one. Suppose $a_1 \in A$, $a_2 \in A$, and $(g \circ f)(a_1) = (g \circ f)(a_2)$. We must show that $a_1 = a_2$. Since $(g \circ f)(a_1) = (g \circ f)(a_2)$, $g(f(a_1)) = g(f(a_2))$. Since g is one-to-one, it follows that $f(a_1) = f(a_2)$. Since f is one-to-one, it follows that $a_1 = a_2$.   **q.e.d.**

Exercises

4.1.1. Let $A = \{1, 2, 3\}$ and $B = \{a, b, c, d\}$. Which of the following are functions from A to B?
(a) $\{(1, a), (2, b)\}$

(b) $\{(1, a), (2, b), (3, c)\}$
(c) $\{(1, a), (2, a), (3, a)\}$
(d) $\{(1, a), (2, b), (3, c), (1, d)\}$
(e) $\{(1, 1), (2, 2), (3, 3)\}$
(f) $\{(1, a), (1, b), (1, c), (1, d)\}$

4.1.2. How many different functions are there from the set $A = \{1, 2, 3\}$ to the set $B = \{\alpha, \beta\}$? List them all.

4.1.3. If A is a set with n elements and B is a set with m elements, how many functions are there from A to B?

4.1.4. If A is a set with n elements, how many bijections are there from A to A?

4.1.5. Suppose A is a set with 4 elements and B is a set with 7 elements.
(a) How many injections are there from A to B?
(b) How many surjections are there from A to B?

4.1.6. Let $f : \mathbb{Z} \to \mathbb{Z}$ be defined by $f(x) = 3x$.
(a) Prove or disprove that f is one-to-one.
(b) Prove or disprove that f is onto.

4.1.7. Let $f : \mathbb{R} \to \mathbb{R}$ be defined by $f(x) = 3x + 5$. Prove that f is a bijection.

4.1.8. Let $f : \mathbb{R} \to \mathbb{R}$ be given by $f(x) = x^5 - 7$. Prove that f is a bijection.

4.1.9. Let $f : \mathbb{N} \times \mathbb{N} \to \mathbb{Q}^+$ be given by $f((n, m)) = n/m$.
(a) Prove or disprove that f is one-to-one.
(b) Prove or disprove that f is onto.

4.1.10. Let $f : A \to B$. Prove that f is a bijection if and only if

$$\forall b \in B \ \exists! a \in A \ \ f(a) = b.$$

4.1.11. Prove: if $f : A \to B$ is a bijection and $g : B \to C$ is a bijection, then $g \circ f$ is a bijection.

4.1.12. Let $f : A \to B$ and $g : B \to C$. Prove or disprove: if $g \circ f$ is onto, then g is onto.

4.1.13. Let $f : A \to B$ and $g : B \to C$. Prove or disprove: if $g \circ f$ is onto, then f is onto.

4.1.14. Let $f : A \to B$ and $g : B \to C$. Prove or disprove: if $g \circ f$ is one-to-one, then g is one-to-one.

4.1.15. Let $f : A \to B$ and $g : B \to C$. Prove or disprove: if $g \circ f$ is one-to-one, then f is one-to-one.

4.1.16. Let $f : A \to B$ and $g : B \to C$. Prove: if $g \circ f$ is onto and g is one-to-one, then f is onto.

4.1.17. Let $f : \mathbb{R} \to \mathcal{P}(\mathbb{R})$ be defined by $f(x) = \{1, 2\} \cup \{x\}$.
(a) Find $f(2)$, $f(3)$, and $f(\pi)$.
(b) Prove or disprove: f is one-to-one.
(c) Prove or disprove: f is onto.

4.1.18. Let $f : \mathbb{R} \to \mathcal{P}(\mathbb{R})$ be defined by $f(x) = (-\infty, x)$.
(a) Explain why f is a function.
(b) Prove or disprove that f is one-to-one.
(c) Prove or disprove that f is onto.

4.1.19. Let A and B be sets, and let f and g be functions from A to B. Define a new function $h : A \to B \times B$ by $h(x) = (f(x), g(x))$.
(a) Prove that if f or g is one-to-one, then h is one-to-one.
(b) Find an example of sets A and B and functions f and g where f and g are both onto but where h is not onto.
(c) Find an example where h is one-to-one but neither f nor g is one-to-one.

4.2   Special functions and inverses

In this section we introduce some functions which are frequently used in different areas of mathematics. Among these are the coordinate projections, the characteristic functions, restrictions of functions, the identity functions, and inverses of functions.

Definition 4.2.1. Let $A_1, A_2, \ldots, A_n$ be nonempty sets. For each index i with $1 \le i \le n$, the function $p_i : A_1 \times A_2 \times \cdots \times A_n \to A_i$ defined by $p_i(a_1, a_2, \ldots, a_n) = a_i$ is called the ith **coordinate projection function**.

Example 4.2.2. Let $A = \{a, b, c, d\}$, $B = \{2, 4, 6, 8, 10\}$, and $C = \mathbb{R}$. Since there are three sets in the cartesian product $A \times B \times C$, there are three coordinate projections: $p_1 : A \times B \times C \to A$, $p_2 : A \times B \times C \to B$, and $p_3 : A \times B \times C \to C$. The projection p_2 is the function which sends $(c, 4, 12.3)$ to 4 and $(a, 10, -6)$ to 10. Similarly, $p_1(d, 2, 0) = d$ and $p_3(c, 8, \pi) = \pi$. The ith coordinate function simply picks out the ith coordinate. □

Example 4.2.3. In calculus there is a theorem (or sometimes a definition) that states that a function $f : \mathbb{R} \to \mathbb{R}^n$ is continuous if and only if for each i with $1 \le i \le n$, the composition $p_i \circ f : \mathbb{R} \to \mathbb{R}$ is continuous. The compositions $p_i \circ f$ are often called the **coordinate functions** of f. For example, the function $f : \mathbb{R} \to \mathbb{R}^3$ given by $f(t) = (t, e^t, \sin t)$ is continuous because each of its coordinate functions $p_1 \circ f(t) = t$, $p_2 \circ f(t) = e^t$, and $p_3 \circ f(t) = \sin t$ is continuous. □

Definition 4.2.4. Let A be a nonempty subset of a universal set U. The **characteristic function on the set** A is the function from U to $\{0, 1\}$ which sends each element of A to the number 1 and each element not in A to the number 0. The characteristic function on A is denoted by χ_A. Thus, $\chi_A : U \to \{0, 1\}$ is given by

$$\chi_A(x) = \begin{cases} 1 & \text{if } x \in A \\ 0 & \text{if } x \notin A \end{cases}$$

Example 4.2.5. Let the universal set be $\mathbb{R}$, and let $A = \mathbb{N}$ and let B be the interval $[-2, 3)$. Then $\chi_A(3) = 1$, $\chi_A(\pi) = 0$, $\chi_B(3) = 0$, and $\chi_B(\sqrt{2}) = 1$.

The characteristic function is also called the **indicator function**. It is used in set theory, real analysis and complex analysis.

Definition 4.2.6. Let $f : X \to Y$ be a function and let $A \subseteq X$. The **restriction of** f **to** A is the function $f|_A : A \to Y$ such that for all $a \in A$, $f|_A(a)$ is defined to be the value $f(a)$.

One difference between a function $f : X \to Y$ and its restriction $f|_A$ is that the domains are different; the domain of f is X and the domain of $f|_A$ is A. Both functions are defined on A, but $f|_A$ is not defined on $X - A$.

Example 4.2.7. Let $f : \mathbb{R} \to \mathbb{R}$ be given by the formula $f(x) = x^2$. Then f is not one-to-one because $f(-1) = f(1)$ but $-1 \ne 1$. Let $A = [0, \infty)$.

Then $f|_A$ is given by the same formula as f, but $f|_A$ is defined only on the nonnegative reals; so $f|_A(-1)$ is undefined. The function $f|_A$ is one-to-one. To prove this, we suppose $a_1 \in A$ and $a_2 \in A$ and $f|_A(a_1) = f|_A(a_2)$. Then $a_1^2 = a_2^2$, and so $\sqrt{a_1^2} = \sqrt{a_2^2}$. Since $a_1 \in A$ and $a_2 \in A$, $a_1 \geq 0$ and $a_2 \geq 0$. Thus, $\sqrt{a_1^2} = a_1$ and $\sqrt{a_2^2} = a_2$. It follows that $a_1 = a_2$, and this finishes a proof that $f|_A$ is one-to-one. The problem in trying the same ideas to attempt to prove that f is one-to-one is that if $x < 0$ then $\sqrt{x^2} \neq x$.   □

Definition 4.2.8. Let A be a nonempty set. Then the **identity function on** A is the function $\mathrm{id}_A : A \to A$ given by

$$\mathrm{id}_A = \{(a, a) : a \in A\}.$$

The identity function on A is the function which sends each element of A to itself. The identity function on $\mathbb{R}$ is given by the formula $f(x) = x$.

Definition 4.2.9. Let $f : A \to B$. Then the function $g : B \to A$ is called an **inverse function** of f if $g \circ f = \mathrm{id}_A$ and $f \circ g = \mathrm{id}_B$. If f has an inverse function then f is called **invertible**.

Example 4.2.10. Let $A = \{1, 2, 3\}$ and $B = \{a, b, c, d\}$. Define $f : A \to B$ by $f = \{(1, a), (2, b), (3, c)\}$. Define $g : B \to A$ by $g = \{(a, 1), (b, 2), (c, 3), (d, 3)\}$. Then $g \circ f = \{(1, 1), (2, 2), (3, 3)\} = \mathrm{id}_A$, but g is not an inverse of f since $f \circ g = \{(a, a), (b, b), (c, c), (d, c)\} \neq \mathrm{id}_B$.   □

Theorem 4.2.11. *Let $f : A \to B$. Suppose g and h are both inverse functions of f. Then $g = h$.*

The proof of Theorem 4.2.11 is Exercise 4.2.6. The theorem asserts that inverses are unique. The inverse function of f is denoted by f^{-1}. Care should be taken when using this notation; for a function f, f^{-1} is *not* the reciprocal of f. Indeed, if the codomain of f is not a set of numbers, the reciprocal is not even defined. In the case when the codomain of f is a set of numbers, the reciprocal can be denoted by $(f(x))^{-1}$. The notation $f^{-1}(x)$ is reserved for the inverse of f applied to the element x.

Theorem 4.2.12. *Let $f : A \to B$. Then f has an inverse function if and only if f is a bijection.*

Proof. ($\Rightarrow$) Suppose f has an inverse function $g : B \to A$. We must show f is one-to-one and onto.

To show f is one-to-one, suppose $f(a_1) = f(a_2)$. Then since g is a function from B to A, $g(f(a_1)) = g(f(a_2))$. Since g is the inverse of f, $g(f(a_1)) = a_1$ and $g(f(a_2)) = a_2$. Hence, $a_1 = a_2$, and so f must be one-to-one.

To show f is onto, suppose $b \in B$. We must show that b is in the range of f; that is, we must find some $a \in A$ such that $f(a) = b$. Since $b \in B$, $g(b) \in A$. Also, since g is the inverse of f, $(f \circ g)(b) = b$. Hence, $f(a) = b$, where $a = g(b)$.

($\Leftarrow$) Next suppose f is a bijection. We must define a function $g : B \to A$ such that $g \circ f = \mathrm{id}_A$ and $f \circ g = \mathrm{id}_B$.

Let $b \in B$ be given. Since f is onto, there must be some $a \in A$ such that $f(a) = b$. Furthermore, since f is one-to-one, there is only one such a. Define $g(b) = a$. By defining g for each given $b \in B$, we have completely defined the function g. It follows from this definition of g that $g \circ f = \mathrm{id}_A$ and $f \circ g = \mathrm{id}_B$.

q.e.d.

Example 4.2.13. It is not difficult to prove that the function $f : (-\infty, 0] \to [0, \infty)$ given by $f(x) = x^2$ is a bijection. By Theorem 4.2.12 this function is invertible. Let's find a formula for f^{-1}. First, the domain of f^{-1} will be the codomain of f, which is $[0, \infty)$. The codomain of f^{-1} will be $(-\infty, 0]$. Now if $f(a) = b$, then $f^{-1}(b) = a$. Starting at the equation $f(a) = b$ with $a \in (-\infty, 0]$ and $b \in [0, \infty)$, if we can solve this equation for a then we will end up with a formula for f^{-1}. In this example, since $f(a) = b$ we get $a^2 = b$. Solving this equation for a without considering where a lies gives us $a = \pm\sqrt{b}$. (It is okay to take the square root of b because $b \in [0, \infty)$.) But since $a \in (-\infty, 0]$, we can omit the positive square root. We get a unique answer: $a = -\sqrt{b}$. This is our formula for f^{-1}: $f^{-1}(b) = -\sqrt{b}$. □

Example 4.2.14. Consider the trigonometric function $f : \mathbb{R} \to \mathbb{R}$ given by $f(x) = \sin x$. Since the range of f is the interval $[-1, 1]$, f is not onto; so by Theorem 4.2.12 f does not have an inverse function. This is easily remedied by defining a new function $g : \mathbb{R} \to [-1, 1]$ given by the same formula $g(x) = \sin x$. The function g is onto; however, g is not one-to-one, so g still does not have an inverse function.

We would like to restrict the domain of g to a longest interval on which g is one-to-one. Indeed, g is one-to-one on the interval $[-\pi/2, \pi/2]$, and it is also one-to-one on the interval $[\pi/2, 3\pi/2]$. The first interval seems more convenient (since it contains 0), and this interval is used in the definition of the arcsine function. Let $h = g|_{[-\pi/2,\pi/2]}$. Then $h : [-\pi/2, \pi/2] \to [-1, 1]$ is a bijection, and it therefore has an inverse. The **arcsine** or **inverse sine** function is the function h^{-1}; it is denoted by $\arcsin x$ or $\sin^{-1} x$.

The sine function is usually considered to be the function f; its domain and codomain are $\mathbb{R}$. Hence, the arcsine function is not precisely the inverse of the sine function; it is the inverse of $\sin|_{[-\pi/2,\pi/2]} : [-\pi/2, \pi/2] \to [-1, 1]$. The composition $(\sin^{-1} \circ \sin)$ is not the identity function on $\mathbb{R}$; for example, $\sin^{-1}(\sin(\pi)) = \sin^{-1}(0) = 0$, not π. The value π is not even in the codomain of the function $\sin^{-1}$. $\square$

Exercises

4.2.1. Find a condition on the sets A and B which is true if and only if both coordinate projections $p_1 : A \times B \to A$ and $p_2 : A \times B \to B$ are one-to-one.

4.2.2. Suppose A and B are subsets of a universal set U. Prove the following properties of the characteristic functions.
(a) $\chi_{A \cap B} = \chi_A \cdot \chi_B$.
(b) $\chi_{A \cup B} = \chi_A + \chi_B - \chi_{A \cap B}$.
(c) $\chi_{\overline{A}} = 1 - \chi_A$.
(d) $\chi_{A \triangle B} = \chi_A + \chi_B - 2\chi_{A \cap B}$, where $A \triangle B$ denotes the symmetric difference of A and B (see page 78).

4.2.3. Let $f : X \to Y$ and $A \subseteq X$. Prove that if f is one-to-one then $f|_A$ is one-to-one.

4.2.4. Let $f : X \to Y$ and $A \subseteq X$. Prove that if $f|_A : A \to Y$ is onto then f is onto.

4.2.5. Construct a specific example of a function $f : X \to Y$ and a subset $A \subseteq X$ such that f is onto but $f|_A$ is not.

4.2.6. Prove Theorem 4.2.11.

4.2.7. Let $f : [0, \infty) \to (-\infty, 0]$ be given by $f(x) = -x^2$. Find a formula for f^{-1}.

4.2.8. Let $f : \mathbb{R} \to \mathbb{R}$ be given by $f(x) = 2x^5 + 7$. Find a formula for f^{-1}.

4.2.9. Let $f : \mathbb{R} \to [-1, 1]$ be given by $f(x) = \cos(x)$. To which intervals can we restrict the domain of f so that the restriction has an inverse function? Which interval is usually used to define the arccosine function?

4.2.10. How can we restrict the domain of the secant function so that the restriction has an inverse function? Which restriction is usually used to define the inverse secant function? Can you find sources that use different restrictions?

4.2.11. Define the function f from the positive integers to $\mathbb{Z}$ by

$$f(n) = \begin{cases} n/2 & \text{if } n \text{ is even} \\ (1 - n)/2 & \text{if } n \text{ is odd}. \end{cases}$$

Does f have an inverse function? If so, find a formula for f^{-1}; if not, prove it does not.

4.2.12. Prove that if $f : A \to B$ is bijective, then $f^{-1} : B \to A$ is a bijection.

4.2.13. Prove that if $f : A \to B$ and $g : B \to C$ are bijections, then $(g \circ f)^{-1} = f^{-1} \circ g^{-1}$.

4.3 Images and inverse images of sets

When f is a function from A to B, it is often important to consider what f does to a whole collection of elements from A.

Definition 4.3.1. Let $f : A \to B$, $S \subseteq A$, and $T \subseteq B$. Then the **image of** S **under** f is the set

$$f(S) = \{b \in B : b = f(s) \text{ for some } s \in S\}.$$

The **inverse image of** T **under** f is the set

$$f^{-1}(T) = \{a \in A : f(a) \in T\}.$$

This is also called the **preimage of** T **under** f.

Given $f : A \to B$, $S \subseteq A$, and $T \subseteq B$, $f(S)$ is the subset of B containing those elements which are output by f when elements of S are input. Likewise, $f^{-1}(T)$ is the subset of A consisting of those elements which f sends into T. Note that $f^{-1}(T)$ is defined even when f is not one-to-one or onto. That is, $f^{-1}(T)$ *is defined even when f has no inverse function.*

Example 4.3.2. Let $A = \{1, 2, 3, 4, 5, 6\}$, $B = \{a, b, c, d, e\}$, and let $f : A \to B$ be given by $f = \{(1, a), (2, a), (3, b), (4, b), (5, c), (6, d)\}$. Furthermore, let $S = \{1, 2, 5\}$ and $T = \{b, c, e\}$. Then $f(S) = \{a, c\}$ since $f(1) = a$, $f(2) = a$ and $f(5) = c$. Also, $f^{-1}(T) = \{3, 4, 5\}$ since $f(3)$, $f(4)$, and $f(5)$ are all elements in T and none of $f(1)$, $f(2)$, and $f(6)$ are in T. □

Example 4.3.3. Consider the function $f : [-3, 3] \to [-2, 2]$ whose graph is shown in Figure 4.3. Find the sets $f([0, 1])$, $f^{-1}([0, 1])$, $f((1, 2))$, and

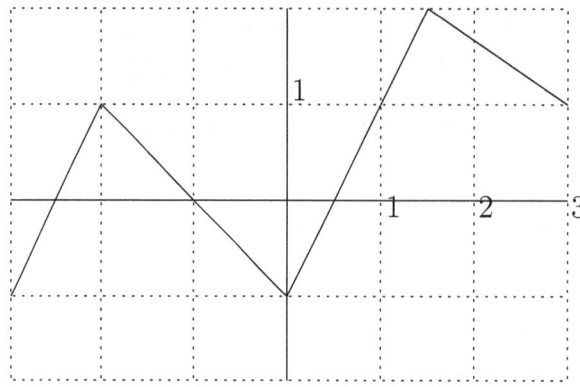

Figure 4.3: A graph of a function $f : [-3, 3] \to [-2, 2]$.

$f^{-1}([-2, -1.5])$.

Solution. The set $f([0, 1])$ is the image of the closed interval $[0, 1]$. This is the set of points $f(x)$ for which $0 \leq x \leq 1$. If we shade in a vertical strip representing the interval $0 \leq x \leq 1$ as in Figure 4.4, then $f([0, 1])$ is the set of all second coordinates of points in the shaded region. Hence, $f([0, 1])$ is the closed interval $[-1, 1]$.

The set $f^{-1}([0, 1])$ is the inverse image of the closed interval $[0, 1]$. This time we shade a horizontal strip representing the interval $0 \leq y \leq 1$ as in Figure 4.5. The set $f^{-1}([0, 1])$ is the set of all first coordinates of points in the shaded region. Hence, $f^{-1}([0, 1]) = [-2.5, -1] \cup [0.5, 1] \cup \{3\}$.

Similarly, $f((1, 2)) = (1, 2]$. Note that $2 \in f((1, 2))$ because $2 = f(1.5)$ and $1.5 \in (1, 2)$. Finally, $f^{-1}([-2, -1.5]) = \emptyset$ because f does not take any points into the horizontal strip $\{y : -2 \leq y \leq -1.5\}$.

□

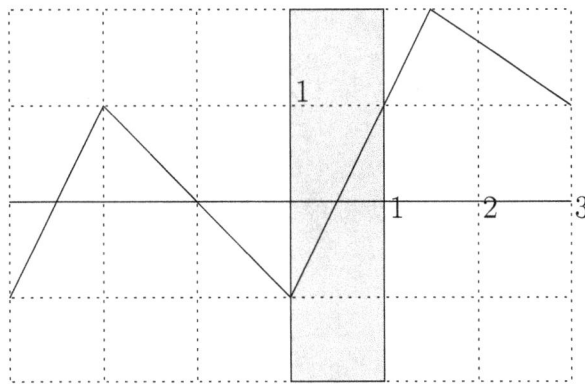

Figure 4.4: $f([0,1]) = \{y : y = f(x) \text{ for some } x \in [0,1]\}$.

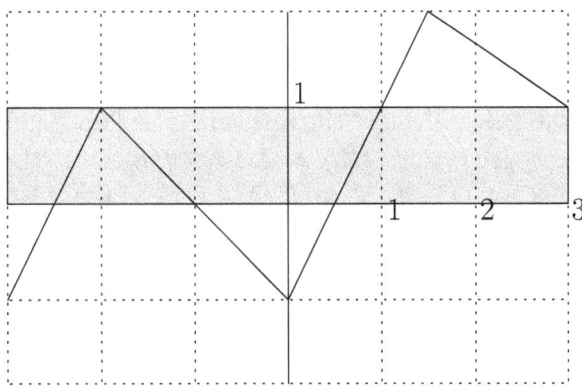

Figure 4.5: $f^{-1}([0,1]) = \{x : f(x) \in [0,1]\}$.

Example 4.3.4. Let S denote the circle $\{(x, y) : x^2 + (y - 1)^2 = 1\}$ in the plane $\mathbb{R} \times \mathbb{R}$, and let $p_1 : \mathbb{R} \times \mathbb{R} \to \mathbb{R}$ and $p_2 : \mathbb{R} \times \mathbb{R} \to \mathbb{R}$ be the coordinate projections. Then the image of S under p_1 is the interval $[-1, 1]$, whereas the image of S under p_2 is the interval $[0, 2]$. The inverse image $p_1^{-1}([-1, 1])$ is the vertical strip $\{(x, y) : -1 \le x \le 1\}$, whereas the inverse image $p_2^{-1}([0, 2])$ is the horizontal strip $\{(x, y) : 0 \le y \le 2\}$. The intersection $p_1^{-1}([-1, 1]) \cap p_2^{-1}([0, 2])$ is the square region $\{(x, y) : -1 \le x \le 1, \ 0 \le y \le 2\}$. Thus, the shape of the circle cannot be reconstructed with knowledge of only its projections on the x- and y-axes.   □

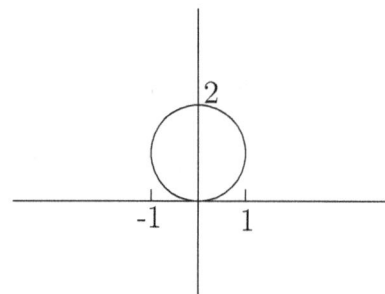

Figure 4.6: The circle $\{(x, y) : x^2 + (y - 1)^2 = 1\}$

Example 4.3.5. Let $f : A \to B$, and let C and D be subsets of A. Is it always true that $f(C - D) = f(C) - f(D)$?

Solution. A simple example will show the answer to be "no." Let $A = \{a, b, c\}$ and $B = \{1, 2, 3\}$. Let $f(a) = 1$, $f(b) = 2$, and $f(c) = 1$. Let $C = \{a, c\}$ and $D = \{b, c\}$. Then $C - D = \{a\}$, so $f(C - D) = \{1\}$. Also, $f(C) = \{1\}$ and $f(D) = \{1, 2\}$, so $f(C) - f(D) = \emptyset$. Since $f(C - D) = \{1\}$ and $f(C) - f(D) = \emptyset$, $f(C - D) \ne f(C) - f(D)$.

However, our counterexample consisted of a function that is not one-to-one. In fact, if we further assume that f is one-to-one, then $f(C - D) = f(C) - f(D)$. Let's give a proof of this claim. Suppose f is one-to-one. In order to show $f(C - D) = f(C) - f(D)$, we will use the pick-a-point method to show $f(C - D) \subseteq f(C) - f(D)$ and $f(C) - f(D) \subseteq f(C - D)$.

First suppose $b \in f(C - D)$. (We are calling this point b because it is a point in the set B.) Then $b = f(a)$ for some $a \in C - D$. Since $a \in C - D$, $a \in C$ and $a \notin D$. Since $b = f(a)$ and $a \in C$, $b \in f(C)$. To show that

$b \notin f(D)$, suppose $b \in f(D)$. Then $b = f(e)$ for some $e \in D$. Then $f(a) = f(e)$, and since f is one-to-one, $a = e$. This contradicts the statement $a \notin D$. By this contradiction, $b \notin f(D)$. Finally, since $b \in f(C)$ and $b \notin f(D)$, $b \in f(C) - f(D)$. Thus, $f(C - D) \subseteq f(C) - f(D)$.

Next suppose $b \in f(C) - f(D)$. Then $b \in f(C)$ and $b \notin f(D)$. Since $b \in f(C)$, $b = f(c)$ for some $c \in C$. Since $b \notin f(D)$ and $b = f(c)$, $c \notin D$. Hence, $c \in C - D$, so $b \in f(C-D)$. This shows that $f(C) - f(D) \subseteq f(C-D)$.

Note that in our proof that $f(C) - f(D) \subseteq f(C - D)$, we never used the supposition that f is one-to-one. Hence, the statement $f(C) - f(D) \subseteq f(C - D)$ is true even when f is not one-to-one. However, in order to make sure the statement $f(C - D) \subseteq f(C) - f(D)$ holds, we do require f to be one-to-one.

$\square$

Exercises

4.3.1. Let $A = \{-2, -1, 0, 1, 2\}$ and $C = \{1, 2\}$. Let $f : A \to \mathbb{R}$ be defined by the formula $f(x) = x^2$.
(a) Find $f(C)$.
(b) Find $f^{-1}(f(C))$.
(c) Does f have an inverse function? Why or why not?
(d) Find $f^{-1}(E)$, where E is the interval $(-\infty, 0)$.

4.3.2. Let $f : \mathbb{R} \to \mathbb{R}$ be defined by $f(x) = x^2$.
(a) Find $f(C)$ where C is the interval $(-3, 2)$.
(b) Find $f^{-1}(E)$ where E is the interval $[-25, 16)$.
(c) Does f have an inverse function? Why or why not?

4.3.3. Let $\mathcal{C}([0, 2])$ denote the set of all continuous functions $f : [0, 2] \to \mathbb{R}$. Let $g : \mathcal{C}([0, 2]) \to \mathbb{R}$ be given by the formula

$$g(f) = \int_0^2 f(x)\, dx.$$

(a) Prove or disprove that g is one-to-one.
(b) Prove or disprove that g is onto.
(c) Find $g(S)$, where $S = \{\sin x, \cos x, e^x\}$.
(d) Prove that $f(x) = 1 - x \in g^{-1}([-1, 3])$.

4.3.4. Let $f : \mathbb{R} \to \mathbb{R}$ be given by $f(x) = x^3 - 4x = x(x-2)(x+2)$. Find $f(C)$ where C is the interval $(0, 2]$.

4.3.5. Let $f : X \to Y$, $A \subseteq X$, and $B \subseteq Y$. Prove that $f|_A^{-1}(B) = A \cap f^{-1}(B)$.

4.3.6. Let $f : A \to B$, $C \subseteq A$, and $D \subseteq A$. Prove that if $C \subseteq D$ then $f(C) \subseteq f(D)$.

4.3.7. Let $f : A \to B$, $E \subseteq B$, and $F \subseteq B$. Prove that if $E \subseteq F$ then $f^{-1}(E) \subseteq f^{-1}(F)$.

4.3.8. Let $f : A \to B$, $C \subseteq A$, and $D \subseteq A$. Prove that $f(C \cap D) \subseteq f(C) \cap f(D)$.

4.3.9. Let $f : A \to B$, $C \subseteq A$, and $D \subseteq A$. Prove by giving a counterexample that it is not always true that $f(C) \cap f(D) \subseteq f(C \cap D)$.

4.3.10. Let $f : A \to B$, $C \subseteq A$, and $D \subseteq A$. Prove that if f is one-to-one then $f(C) \cap f(D) \subseteq f(C \cap D)$.

4.3.11. Let $f : A \to B$, $E \subseteq B$, and $F \subseteq B$. Prove that $f^{-1}(E \cap F) = f^{-1}(E) \cap f^{-1}(F)$.

4.3.12. Let $f : A \to B$, $C \subseteq A$, and $D \subseteq A$. Prove that $f(C \cup D) = f(C) \cup f(D)$.

4.3.13. Let $f : A \to B$, $E \subseteq B$, and $F \subseteq B$. Prove that $f^{-1}(E \cup F) = f^{-1}(E) \cup f^{-1}(F)$.

4.3.14. Let $f : A \to B$ and $E \subseteq B$. Prove that $f(f^{-1}(E)) \subseteq E$.

4.3.15. Let $f : A \to B$ and $E \subseteq B$. Prove by giving a counterexample that it is not always true that $f(f^{-1}(E)) = E$.

4.3.16. Let $f : A \to B$ and $E \subseteq B$. Prove that if f is onto then $f(f^{-1}(E)) = E$.

4.3.17. Let $f : A \to B$ and $C \subseteq A$. Prove that $C \subseteq f^{-1}(f(C))$.

4.3.18. Let $f : A \to B$ and $C \subseteq A$. Prove that it is not always true that $C = f^{-1}(f(C))$.

4.3.19. Let $f : A \to B$ and $C \subseteq A$. Prove that if f is one-to-one then $C = f^{-1}(f(C))$.

4.4 Arbitrary unions and intersections

Whenever we are working with a large number of sets, it is convenient to use indices to name the sets. For example, if there are 10 sets under consideration, we might name them A_1, A_2, A_3, and so forth, up to A_{10}. The indices do not have to be integers. It might be convenient to use real numbers for indices, like $A_{1/2}$ or $A_{\sqrt{3}}$.

Definition 4.4.1. Let Λ denote any set, and for each $\lambda \in \Lambda$, let A_λ denote a set. Then the collection of sets $\{A_\lambda\}_{\lambda \in \Lambda}$ is called an **indexed collection of sets** or an **indexed family of sets**, and the set Λ is called the **index set**.

Once an indexed collection of sets is specified, we can define the union and intersection of all of the sets in the collection:

Definition 4.4.2. If $\{A_\lambda\}_{\lambda \in \Lambda}$ is an indexed collection of sets, then the union and intersection of all of the A_λ are defined by

$$\bigcup_{\lambda \in \Lambda} A_\lambda = \{x : x \in A_\lambda \text{ for some } \lambda \in \Lambda\}$$

and

$$\bigcap_{\lambda \in \Lambda} A_\lambda = \{x : x \in A_\lambda \text{ for every } \lambda \in \Lambda\}.$$

If $\Lambda = \{1, 2, 3, ..., k\}$, then $\bigcup_{\lambda \in \Lambda} A_\lambda$ is also denoted by

$$\bigcup_{n=1}^{k} A_n.$$

Likewise, if Λ is the set of all positive integers, then $\bigcup_{\lambda \in \Lambda} A_\lambda$ is also denoted by

$$\bigcup_{n=1}^{\infty} A_n.$$

The examples in this section make use of **interval notation** for intervals in the real line. Given real numbers a and b with $a < b$, the intervals (a, b),

$[a, b)$, and so forth are defined as follows:

$$\begin{aligned}
(a, b) &= \{x \in \mathbb{R} : a < x < b\} \\
[a, b] &= \{x \in \mathbb{R} : a \le x \le b\} \\
[a, b) &= \{x \in \mathbb{R} : a \le x < b\} \\
(a, b] &= \{x \in \mathbb{R} : a < x \le b\} \\
(a, \infty) &= \{x \in \mathbb{R} : a < x\} \\
(-\infty, b) &= \{x \in \mathbb{R} : x < b\} \\
[a, \infty) &= \{x \in \mathbb{R} : a \le x\} \\
(-\infty, b] &= \{x \in \mathbb{R} : x \le b\}
\end{aligned}$$

Care should be taken when using the notation (a, b), for this is used to denote both an interval as above and a point in the plane. Hopefully which concept is intended will be clear from the context.

Example 4.4.3. Let $\Lambda = \mathbb{R}^+$ and for each $\lambda \in \mathbb{R}^+$, let A_λ denote the interval $[0, 2 + \lambda)$. Then $\bigcup_{\lambda \in \mathbb{R}^+} A_\lambda$ is the union of all of the intervals $[0, 2 + \lambda)$ where λ ranges over all positive numbers. Thus, $\bigcup_{\lambda \in \mathbb{R}^+} A_\lambda = [0, \infty)$. Also, $\bigcap_{\lambda \in \mathbb{R}^+} A_\lambda$ is the intersection of all of the intervals $[0, 2 + \lambda)$, and this intersection is equal to the single interval $[0, 2]$. Note that we include 2 in the intersection because 2 is an element of every one of the intervals A_λ. $\square$

Example 4.4.4. Let $\Lambda = \mathbb{N}$ and for each $\lambda \in \mathbb{N}$, let A_λ denote the interval $(-\lambda, \lambda + 1)$. Then $\bigcup_{\lambda \in \mathbb{N}} A_\lambda = \mathbb{R}$ and $\bigcap_{\lambda \in \mathbb{N}} A_\lambda = (-1, 2)$. $\square$

Example 4.4.5. For each positive integer n, let $A_n = (-2 - 1/n, 3 + 1/n]$. Then

$$\bigcup_{n=1}^{\infty} A_n = A_1 \cup A_2 \cup \cdots = (-3, 4] \cup (-2.5, 3.5] \cup \cdots = (-3, 4]$$

and

$$\bigcap_{n=1}^{\infty} A_n = A_1 \cap A_2 \cap \cdots = (-3, 4] \cap (-2.5, 3.5] \cap \cdots = [-2, 3].$$

To prove either of these equalities, we can use the method of showing that each is a subset of the other. For example, to prove that

$$\bigcup_{n=1}^{\infty} A_n \subseteq (-3, 4],$$

suppose $x \in \bigcup_{n=1}^{\infty} A_n$. Then $x \in A_j$ for some positive integer j. Then

$$-2 - 1/j < x \le 3 + 1/j.$$

Since $j \ge 1$,

$$-3 \le -2 - 1/j$$

and

$$3 + 1/j \le 4.$$

Hence, $-3 < x \le 4$, so $x \in (-3, 4]$.

The proof that $(-3, 4] \subseteq \bigcup_{n=1}^{\infty} A_n$ follows easily from the fact that $A_1 = (-3, 4]$.

To prove $\bigcap_{n=1}^{\infty} A_n \subseteq [-2, 3]$, suppose $x \in \bigcap_{n=1}^{\infty} A_n$. Then for every positive integer n, $-2 - 1/n < x \le 3 + 1/n$. Since we can make $1/n$ arbitrarily close to 0 (but not equal to 0), $-2 \le x \le 3$. So $x \in [-2, 3]$.

Finally, to prove $[-2, 3] \subseteq \bigcap_{n=1}^{\infty} A_n$, suppose $x \in [-2, 3]$. Then $-2 \le x \le 3$. Now for every positive integer n, $-2 - 1/n < -2$ and $3 < 3 + 1/n$. Hence $x \in (-2 - 1/n, 3 + 1/n)$ for every n. Therefore, $x \in A_n = (-2 - 1/n, 3 + 1/n]$ for every positive integer n and $x \in \bigcap_{n=1}^{\infty} A_n$. □

Example 4.4.6. Find $\bigcup_{n=1}^{\infty} \left[0, 4 - \dfrac{1}{n} \right]$ and prove the result.

Solution. We might initially compute the first few terms in the union:

$$\bigcup_{n=1}^{\infty} \left[0, 4 - \frac{1}{n} \right] = [0, 3] \cup \left[0, 3\frac{1}{2} \right] \cup \left[0, 3\frac{2}{3} \right] \cup \left[0, 3\frac{3}{4} \right] \cup \cdots .$$

It seems the union must equal either $[0, 4]$ or $[0, 4)$, so the question is whether 4 is in the union or not. The limit of the expression $4 - \frac{1}{n}$ as $n \to \infty$ is exactly 4, but the number 4 is actually in *none* of the intervals of the form $[0, 4 - \frac{1}{n}]$. In order to be in the union, 4 must appear in at least one of the sets in the collection. Hence, 4 is not in the union and the union must be $[0, 4)$.

To prove $\bigcup_{n=1}^{\infty} \left[0, 4 - \frac{1}{n}\right] = [0, 4)$, we prove $\bigcup_{n=1}^{\infty} \left[0, 4 - \frac{1}{n}\right] \subseteq [0, 4)$ and $[0, 4) \subseteq \bigcup_{n=1}^{\infty} \left[0, 4 - \frac{1}{n}\right]$.

To prove $\bigcup_{n=1}^{\infty} \left[0, 4 - \frac{1}{n}\right] \subseteq [0, 4)$, suppose $x \in \bigcup_{n=1}^{\infty} \left[0, 4 - \frac{1}{n}\right]$. Then for some positive integer j, $x \in [0, 4 - \frac{1}{j}]$. Thus, $0 \le x \le 4 - \frac{1}{j}$. Since $4 - \frac{1}{j} < 4$, $0 \le x < 4$. Hence, $x \in [0, 4)$.

To prove $[0, 4) \subseteq \bigcup_{n=1}^{\infty} \left[0, 4 - \frac{1}{n}\right]$, suppose $x \in [0, 4)$. Then $0 \le x < 4$. Since $x < 4$, there must be some (possibly large) integer j such that $x \le 4 - \frac{1}{j}$. Then $0 \le x \le 4 - \frac{1}{j}$, so $x \in [0, 4 - \frac{1}{j}]$. Since x is in this particular set, x must be in the union $\bigcup_{n=1}^{\infty} \left[0, 4 - \frac{1}{n}\right]$. $\square$

Definition 4.4.7. Let Γ be any collection of sets. Then the union and intersection over all sets in Γ are defined by

$$\bigcup \Gamma = \{x : x \in A \text{ for some } A \in \Gamma\}$$

and

$$\bigcap \Gamma = \{x : x \in A \text{ for every } A \in \Gamma\}.$$

Example 4.4.8. Let $\Gamma = \{(0, 3), (1, 4), [2, 5]\}$. Then

$$\bigcup \Gamma = (0, 3) \cup (1, 4) \cup [2, 5] = (0, 5]$$

and

$$\bigcap \Gamma = (0, 3) \cap (1, 4) \cap [2, 5] = [2, 3).$$

$\square$

Example 4.4.9. Prove that if $\Gamma \subseteq \Omega$, then $\bigcup \Gamma \subseteq \bigcup \Omega$.

Solution. Before jumping right into a proof, it is a good idea to do examples if possible to clarify the statement to be proven. Since we are dealing with the union $\bigcup \Gamma$, Γ must be a collection of sets. Similarly, Ω must also be a collection of sets. For an example where $\Gamma \subseteq \Omega$, let's just let

$$\Gamma = \{\ \{1,2\}, \{2,3\}, \{3,4\}\ \}$$

and

$$\Omega = \{\ \{1,2\}, \{2,3\}, \{3,4\}, \{4,5\}\ \}.$$

Then $\bigcup \Gamma = \{1,2,3,4\}$ and $\bigcup \Omega = \{1,2,3,4,5\}$. Notice that the conclusion $\bigcup \Gamma \subseteq \bigcup \Omega$ is true for our example. In this example if we supposed that $x \in \bigcup \Gamma$, then x would be one of the numbers 1, 2, 3, or 4. Each of these numbers is in at least one of the sets in the collection Γ. Here's a proof.

Proof. Suppose $\Gamma \subseteq \Omega$, and suppose $x \in \bigcup \Gamma$. Then by Definition 4.4.2, $x \in A$ for some set $A \in \Gamma$. Since $\Gamma \subseteq \Omega$ and $A \in \Gamma$, $A \in \Omega$ too. Since $x \in A$ and $A \in \Omega$, $x \in \bigcup \Omega$. **q.e.d.**

$\square$

Exercises

4.4.1. Let $\Lambda = \mathbb{Z}$ and for each $\lambda \in \mathbb{Z}$, let $A_\lambda = [\lambda, \lambda + 2]$. Find $\bigcup_{\lambda \in \mathbb{Z}} A_\lambda$ and $\bigcap_{\lambda \in \mathbb{Z}} A_\lambda$.

4.4.2. Let $\Gamma = \{[0,3), (2,4], (1,4)\}$. Find $\bigcap \Gamma$ and $\bigcup \Gamma$.

4.4.3. Let $\Omega = \{\ \{\alpha, \beta, \gamma, \delta\}, \{\alpha, \eta, \delta\}, \{\alpha, \zeta, \varepsilon, \delta, \sigma\}\ \}$. Find $\bigcap \Omega$ and $\bigcup \Omega$.

4.4.4. For any integer n, let M_n denote the set of multiples of n. (For example, M_2 is the set of even integers.)
(a) Find $M_6 \cap M_8$.
(b) Find $\bigcap_{n \in \Lambda} M_n$, where $\Lambda = \{6, 8, 10\}$.

(c) Find $\bigcup_{n \in \Lambda} M_n$, where Λ is the set of all primes.

(d) Find $\bigcap_{n \in \Lambda} M_n$, where Λ is the set of all primes.

4.4.5. Find $\bigcup_{n=1}^{\infty}(-1/n, 1/n)$ and prove your answer.

4.4.6. Find $\bigcap_{n=1}^{\infty}(-1/n, 1/n)$ and prove your answer.

4.4.7. Find $\bigcap_{n=1}^{\infty}(0, 1/n)$ and prove your answer.

4.4.8. Find $\bigcup_{n=1}^{\infty}\left[3+\frac{1}{n},\ 6-\frac{1}{n}\right]$ and prove your answer.

4.4.9. Let $\Omega = \{A_n : n \in \mathbb{Z}\}$, where A_n is the interval $(n-1, n+1)$.
(a) Find $\bigcap \Omega$.
(b) Find $\bigcup \Omega$.
(c) Is $(-1, 1) \in \Omega$?
(d) Is $(-1, 1) \in \bigcup \Omega$?
(e) Is $0 \in \Omega$?
(f) Is $0 \in \bigcup \Omega$?

4.4.10. Prove or disprove: if $\bigcup \Gamma = \bigcup \Omega$, then $\Gamma = \Omega$.

4.4.11. Prove or disprove: if $\bigcap \Gamma = \bigcap \Omega$, then $\Gamma = \Omega$.

4.4.12. Prove or disprove: if $\Gamma \subseteq \Omega$, then $\bigcap \Omega \subseteq \bigcap \Gamma$.

4.4.13. Find $\bigcup \emptyset$ and $\bigcap \emptyset$.

4.4.14. Prove DeMorgan's Laws for arbitrary unions and intersections of sets:
(a) $\overline{\left(\bigcup \Gamma\right)} = \bigcap \{\overline{A} : A \in \Gamma\}$
(b) $\overline{\left(\bigcap \Gamma\right)} = \bigcup \{\overline{A} : A \in \Gamma\}$

4.4.15. Prove the distributive law
$$B \cap \left(\bigcup \Gamma\right) = \bigcup \{B \cap A : A \in \Gamma\}.$$

4.5 Cardinality

In Section 3.1 we defined the *cardinality* of a set to be the number of elements in the set. Think back to when you first learned how to count, and consider the set $S = \{$blue, green, red, yellow$\}$. How would we determine the number of elements in this set? By counting, of course. We can proceed to count the items one at a time, counting out each number as we point at the next color in the set. We might start by saying *one* when we point at the color *blue*, and then count out *two, three, four* when we point at the colors *green, red*, and *yellow*, in that order.

It would not be incorrect to count the colors in a different order; by saying *one* when we point at *red*, and following this with *two, three, four* when we point at *green, yellow*, and *blue*, in that order, we would again come to the conclusion that there are four colors in this set.

However, when you were first learning how to count, you undoubtedly made a few mistakes. I saw this first-hand when my daughter Kelsey was one year old. When counting the colors in a painting, she might have said *one* and pointed at *blue*, and proceeded to count *two* for *red, three* for *green, four* for *yellow*, and *five* for *blue*. That is, she might have counted the same color twice. On other occasions, she might have missed a color completely.

Counting the elements of the set $S = \{$blue, green, red, yellow$\}$ is nothing more than defining a bijection $f : \{1, 2, 3, 4\} \to S$. By pointing at a color and calling out one of the integers in $\{1, 2, 3, 4\}$ we are defining the function. For example, if you point at *red* first and say *one*, then $f(1) = $ red. If you count the same color twice, then the function is not one-to-one. If you miss a color completely, then the function is not onto.

Hence, given a set S, $|S| = n$ if and only if there is a bijection $f : \{1, 2, 3, \ldots, n\} \to S$. We would like to extend this idea to infinite sets; when we do so we will find out that there are different kinds of infinity.

Definition 4.5.1. Two sets A and B are said to **have the same cardinality** if there is a bijection $f : A \to B$. The set A is said to **have cardinality greater than that of B** if there is no injective function $f : A \to B$. The set A is said to **have cardinality less than that of B** if there is no surjective function $f : A \to B$. A set S is called **finite** if $|S| = n$ for some integer n; otherwise, S is called **infinite**. The set S is called **countable** if it is finite or if there is a bijection $f : \mathbb{N} \to S$. A set is called **uncountable** if it is not countable.

Example 4.5.2. Is the set of all integers a countable set?

Solution. At first glance it might look like there is no bijection $f : \mathbb{N} \to \mathbb{Z}$, since $\mathbb{N}$ is a proper subset of $\mathbb{Z}$. However, there is indeed such a bijection. To see an example of a bijection $f : \mathbb{N} \to \mathbb{Z}$, look at the set of all integers on a real number line as in Figure 4.7. Then simply start somewhere, and devise a method of counting all the integers. But you must be sure that, given any integer m, this integer m is actually counted at some point. As Figure 4.7 demonstrates, we can start with 0 and then alternate between positive and negative integers.

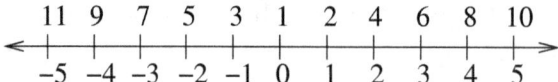

Figure 4.7: A bijection $f : \mathbb{N} \to \mathbb{Z}$.

The bijection constructed in Figure 4.7 is given by the formula

$$f(n) = \begin{cases} n/2 & \text{if } n \text{ is even} \\ (1-n)/2 & \text{if } n \text{ is odd} \end{cases}$$

By Exercise 4.2.11, f has an inverse function. By Theorem 4.2.12, f must be a bijection. Hence, the set $\mathbb{Z}$ is countable.   □

Example 4.5.2 illustrates that the sets $\mathbb{N}$ and $\mathbb{Z}$ have the same cardinality. Even though $\mathbb{N}$ is a proper subset of $\mathbb{Z}$, these two sets really have the same number of elements. This number of elements is denoted by $\aleph_0$; that is, $|\mathbb{N}| = \aleph_0$. (The symbol $\aleph$ is the first letter of the Hebrew alphabet, "aleph.") The following proposition verifies that not all infinite sets have the same cardinality.

Proposition 4.5.3. *The interval* $(0, 1)$ *is uncountable.*

Proof. The following proof is known as *Cantor's diagonalization argument*, named after the famous mathematician Georg Cantor (1845–1918).

Suppose $(0, 1)$ is countable. Then there exists a bijection $f : \mathbb{N} \to (0, 1)$. Now for each $j \in \mathbb{N}$, write $f(j)$ in decimal form as $f(j) = 0.d_{j1}d_{j2}d_{j3}d_{j4}\ldots$, where d_{jk} is the kth digit in a decimal representation of the real number $f(j)$. For example, if $f(1) = 1/3$, then $d_{1k} = 3$ for all k.

We will show that f cannot be onto because there is a real number in $(0,1)$ which is not equal to any $f(j)$. This contradiction will finish the proof that $(0,1)$ must be uncountable.

Define the decimal number $c = 0.b_1 b_2 b_3 \cdots$ by

$$b_\ell = \begin{cases} 1 & \text{if } d_{\ell\ell} = 2 \\ 2 & \text{if } d_{\ell\ell} \neq 2 \end{cases}.$$

Then c cannot equal $f(1)$ because the decimal expansions of c and $f(1)$ differ in the first decimal: $b_1 \neq d_{11}$. Here we use the fact that the decimal expansion of a real number is unique except when the expansion ends in all 0's or all 9's, as in the equality $0.249999\cdots = 0.250000\cdots$. The decimal expansion of c contains only 1's and 2's so this exception is not problematic here. In general, c cannot equal $f(j)$ for any j because c and $f(j)$ differ in the jth decimal: $b_j \neq d_{jj}$. Thus, $c \in (0,1)$ but c is not in the range of f. So f is not onto.   **q.e.d.**

Theorem 4.5.4. *The set $\mathbb{R}$ is uncountable.*

Theorem 4.5.4 follows from Proposition 4.5.3 and the fact that if a set contains an uncountable subset then the set itself must be uncountable; the proof of this is Exercise 4.5.4. The sets $(0,1)$ and $\mathbb{R}$ do have the same cardinality; the function $f : (0,1) \to \mathbb{R}$ defined by $f(x) = \tan(\pi(x - 1/2))$ is a bijection.

Theorem 4.5.5. *Let S be a nonempty set. Then the following are equivalent:*
(1) S is countable;
(2) there is a one-to-one function $f : S \to \mathbb{N}$; and
(3) there is an onto function $f : \mathbb{N} \to S$.

The meaning of the claim *The following are equivalent* in Theorem 4.5.5 is that statement (1) is true if and only if (2) is true; and (2) is true if and only if (3) is true; and thus (1) is true if and only if (3) is true. To prove such a theorem it would suffice to prove if (1) then (2); if (2) then (3); and if (3) then (1).

Proposition 4.5.6. *The Cartesian product $\mathbb{N} \times \mathbb{N}$ is countable.*

Proof. We will use Theorem 4.5.5 and define a function $f : \mathbb{N} \times \mathbb{N} \to \mathbb{N}$ which is one-to-one. One such function that works is

$$f((n, m)) = 2^n 3^m.$$

To show that this function f is one-to-one, suppose $f((n, m)) = f((k, \ell))$. Then $2^n 3^m = 2^k 3^\ell$. Now if $n > k$, then $2^{n-k} 3^m = 3^\ell$ and so 3^ℓ is even; this is a contradiction. If $k > n$ then $3^m = 2^{k-n} 3^\ell$ and so 3^m is even; this is also a contradiction. Hence, $n = k$ and so $3^m = 3^\ell$, from which it follows that $m = \ell$. Therefore, $(n, m) = (k, \ell)$, so f is one-to-one.   **q.e.d.**

One way to "count" the elements of $\mathbb{N} \times \mathbb{N}$ is in the following order: $(1, 1)$, $(1, 2)$, $(2, 1)$, $(1, 3)$, $(2, 2)$, $(3, 1)$, $(1, 4)$, $(2, 3)$, and so on. Fist we count all those (n, m) such that $n + m = 2$; then all those with $n + m = 3$; and then all those with $n + m = 4$, and so forth.

Proposition 4.5.7. *The set $\mathbb{Q}^+$ of all positive rational numbers is countable.*

Proof. Define $f : \mathbb{N} \times \mathbb{N} \to \mathbb{Q}^+$ by $f((n, m)) = n/m$. Then f is surjective. Since $\mathbb{N} \times \mathbb{N}$ is countable by Proposition 4.5.6, $\mathbb{Q}^+$ must be countable by Exercise 4.5.1.   **q.e.d.**

The set $\mathbb{Q}$ of all rationals is also countable. (See Exercise 4.5.3.)

Proposition 4.5.8. *If A and B are countable sets, then the Cartesian product $A \times B$ is also countable.*

Proof. Suppose A and B are countable. Then by Theorem 4.5.5, there are two onto functions $f : \mathbb{N} \to A$ and $g : \mathbb{N} \to B$. Define the function $h : \mathbb{N} \times \mathbb{N} \to A \times B$ by $h((n, m)) = (f(n), g(m))$. We claim that h is onto; it will follow from Exercise 4.5.1 that $A \times B$ is countable. To prove that h is onto, suppose $(a, b) \in A \times B$. Then $a \in A$ and $b \in B$. Since $a \in A$ and f is onto, there is some $n \in \mathbb{N}$ such that $f(n) = a$. Similarly, since $b \in B$ and g is onto, there is some $m \in \mathbb{N}$ such that $g(m) = b$. Then $h((n, m)) = (a, b)$, so h is onto.   **q.e.d.**

Proposition 4.5.9. *The countable union of countable sets is countable.*

Proof. Let Λ be any countable set, and for each $\lambda \in \Lambda$, let A_λ be a countable set. We must prove that $\bigcup_{\lambda \in \Lambda} A_\lambda$ is countable. To do this, we will construct an onto function $f : \mathbb{N} \times \mathbb{N} \to \bigcup A_\lambda$. It follows from Proposition 4.5.6 and Exercise 4.5.1 that $\bigcup A_\lambda$ is countable.

Since A_λ is countable for each $\lambda \in \Lambda$, for each $\lambda \in \Lambda$ there is an onto function $g_\lambda : \mathbb{N} \to A_\lambda$. Similarly, since Λ is countable, there is an onto function $h : \mathbb{N} \to \Lambda$.

Define $f : \mathbb{N} \times \mathbb{N} \to \bigcup A_\lambda$ by $f((n, m)) = g_{h(n)}(m)$. To show that f is onto, suppose $a \in \bigcup A_\lambda$. Then $a \in A_{\lambda_0}$ for some $\lambda_0 \in \Lambda$. Since h is onto, there is some $n_0 \in \mathbb{N}$ such that $h(n_0) = \lambda_0$. Since g_{λ_0} is onto, there is some $m_0 \in \mathbb{N}$ such that $g_{\lambda_0}(m_0) = a$. Then $f((n_0, m_0)) = a$. **q.e.d.**

Just as the countable union of countable sets is countable, one might suspect that the countable Cartesian product of countable sets is countable. However, Exercise 4.5.5 shows that this is not the case.

Proposition 4.5.10. *Let S be any set. Then S has cardinality less than that of the power set $\mathcal{P}(S)$.*

Proof. Let S be any set. To proceed by contradiction, suppose there is a surjective function $f : S \to \mathcal{P}(S)$.

Consider the set $T = \{s \in S : s \notin f(s)\}$. The set T is a subset of S, and so $T \in \mathcal{P}(S)$. Since f is onto, T must be in the range of f; that is, $f(t) = T$ for some $t \in S$.

Now either $t \in T$ or $t \notin T$. If $t \in T$, then $t \in f(t)$, and by the definition of T, $t \notin T$. This is a contradiction. If $t \notin T$, then $t \notin f(t)$, so $t \in T$ by the definition of T. This is also a contradiction. Either way, we get a contradiction; so there is no surjective function $f : S \to \mathcal{P}(S)$. **q.e.d.**

By Proposition 4.5.10, there are infinitely many different sizes of infinity. The smallest is $\aleph_0$, the cardinality of the natural numbers. One of the axioms of set theory called the **Axiom of Choice** can be used to prove that there is a smallest cardinality $\aleph_1$ which is greater than $\aleph_0$; and similarly there is a whole infinite sequence of different infinite cardinalities: $\aleph_0, \aleph_1, \aleph_2, \ldots$. The axiom of choice says that given an arbitrary collection of nonempty sets it is possible to choose an element from each set. Equivalently, the Cartesian product of arbitrarily many nonempty sets is nonempty.

The **Continuum Hypothesis** is the statement that the cardinality of $\mathbb{R}$ is $\aleph_1$; that is, there is no set with cardinality less than that of $\mathbb{R}$ and greater than that of $\mathbb{N}$. The term *continuum* is another word for the real line $\mathbb{R}$. The Continuum Hypothesis is a hypothesis or an axiom of set theory; it was conjectured by Georg Cantor in 1877. In 1939 Kurt Gödel proved that one could not disprove the Continuum Hypothesis using the usual axioms of set theory, and in 1963 Paul Cohen proved that one could not prove it either. Whether the mathematical community as a whole will generally accept or reject the hypothesis as an axiom seems uncertain at this point in time. One would like all accepted axioms to be as intuitively obvious as possible, but the Continuum Hypothesis is far from intuitively obvious. Perhaps one day someone will find an axiom that seems obvious and that can be used to prove either that the Continuum Hypothesis is true or that it is false. At present there is a lot of interest in this area of mathematical research.

Exercises

4.5.1. Show that if A is countable and there is a surjective function $f : A \to B$, then B is countable.

4.5.2. Show that if A is countable and there is an injective function $f : B \to A$, then B is countable.

4.5.3. Show that the set $\mathbb{Q}$ of all rational numbers is countable.

4.5.4. (a) Show that if there is a one-to-one function $f : A \to B$ and $C \subseteq A$, then there is a one-to-one function $g : C \to B$.
(b) Show that if A is uncountable and $A \subseteq B$, then B is uncountable.
(c) Show that $\mathbb{R}$ is uncountable.
(d) Show that if B is countable and $A \subseteq B$, then A is countable.

4.5.5. For each positive integer n, let $A_n = \{0, 1\}$. Also, let

$$A = \prod_{n=1}^{\infty} A_n = A_1 \times A_2 \times A_3 \times \cdots.$$

Then A can be thought of as the set of all sequences of 0's and 1's. Use Cantor's diagonalization argument (see the proof of Proposition 4.5.3) to prove that A is uncountable.

4.5.6. For each positive integer n, let $A_n = \{0, 1\}$. Also, let

$$A = \prod_{n=1}^{\infty} A_n = A_1 \times A_2 \times A_3 \times \cdots.$$

Then A can be thought of as the set of all sequences of 0's and 1's.
(a) Given any nonnegative integer k, let B_k be the subset of A consisting of all sequences which have exactly k 1's. Show that B_k is countable.

(b) Let B be the subset of A consisting of all sequences which have a finite number of 1's. Show that B is countable.

(c) Let C be the subset of A consisting of all sequences which end in all 1's. Show that C is countable.

4.5.7. A real number is **algebraic** if it is the root of some polynomial with integer coefficients. Otherwise it is a **transcendental** number. For example, $\sqrt{2}$ is algebraic since it is a root of the polynomial $x^2 - 2$.

(a) Prove that the number $\sqrt{2 + \sqrt{3} + \sqrt{5}}$ is an algebraic number by finding a polynomial with integer coefficients for which this number is a root.

(b) Prove that there are only countably many algebraic numbers.

(c) Prove that there are uncountably many transcendental numbers.

4.5.8. Find an infinite sequence of infinite sets $A_1, A_2, A_3, \ldots$ such that for all indices $n \geq 1$, the set A_n has cardinality less than that of A_{n+1}.

Chapter 5

What's next?

A child's ... first geometrical discoveries are topological ... If you ask him to copy a square or a triangle, he draws a closed circle.

-Jean Piaget, pioneer of developmental psychology (1896-1980)

Algebra is generous; she often gives more than is asked of her.

-Jean Le Rond d'Alembert, mathematician (1717-1783)

After your study of the first four chapters in this textbook, you are probably ready to begin a study of mathematical topics that would usually be studied after your first studies of calculus. Three main branches of mathematics you might study at this stage include topology, abstract algebra, and analysis. Of course there are many other topics in mathematics to study, but it would be difficult to get very far in any mathematical field without a good understanding of at least one of these three branches.

Topology is similar to geometry in that the objects of study involve shapes like squares and spheres. Topology differs from geometry in that exact measurements of lengths and angles matter in geometry but such measurements do not matter in topology. In topology, what matters is how a shape is assembled from points into a unified whole.

Abstract algebra is the study of structures put on sets via generalized operations like addition, multiplication, exponentiation, or composition.

Topology and abstract algebra are similar in that structures on sets are studied in both areas. The difference is that the topological structure involves how the points fit together to form a shape, whereas the algebraic structure involves how an operation acts on the points.

Analysis is the generalized study of calculus. Analysis is generally divided into real analysis, which involves functions of real variables, and complex analysis, which involves functions of complex variables.

In this chapter we provide very brief introductions to topology and abstract algebra. Since you probably have some experience with calculus, you might have already tasted the flavor of analysis.

5.1   Topology, continuity and homeomorphisms

Topology is the study of the shape and structure of geometric objects that remain unchanged when the objects are stretched or bent. For example, a circle can easily be stretched and bent to form an ellipse, a triangle, or a square; but no amount of stretching can make a circle take the form of a line or a pair of disjoint circles. Similarly, a sphere can be stretched to look like a cube, but one cannot stretch a sphere so that it looks like a bicycle tire.

The structure of a geometric object is achieved by considering the object to be a set along with a collection of subsets with certain basic properties. The collection of subsets is called a *topology* on the object. The intuitive ideas of stretching and bending one object to make another are made mathematically rigorous by defining a function from the first object to the second object and studying how the function relates the subsets of the two objects.

Today topology is a major branch of mathematics, and it has connections with almost every other branch of mathematics. Applications of topology are found in computer science (networking), biology (DNA structures), economics, physics, chemistry, and many other fields.

Formally, a topology on a set is defined as follows.

Definition 5.1.1. Let X be a set. Then a collection $\mathcal{T}$ of subsets of X is a **topology** on X if the following three conditions are satisfied:

1. $X \in \mathcal{T}$ and $\emptyset \in \mathcal{T}$;

2. if $A \in \mathcal{T}$ and $B \in \mathcal{T}$ then $A \cap B \in \mathcal{T}$; and

3. if $A_\alpha \in \mathcal{T}$ for every α in some index set Λ, then

$$\bigcup_{\alpha \in \Lambda} A_\alpha \in \mathcal{T}.$$

When $\mathcal{T}$ is a topology on X, the elements of $\mathcal{T}$ are called **open sets**, and the pair $(X, \mathcal{T})$ is called a **topological space**. A subset C of X is called **closed** if its complement $X - C$ is an open set.

Example 5.1.2. Let X be any set, and let $\mathcal{D}$ be the power set of X (the collection of all subsets of X). Then since $\mathcal{D}$ satisfies the three properties in Definition 5.1.1, $\mathcal{D}$ is a topology on X. This is called the **discrete topology** on X.

   The set $\{\emptyset, X\}$ is also a topology on X, called the **trivial topology** on X. $\square$

Example 5.1.3. Let X be the set of real numbers $\mathbb{R}$. Then the collection $\mathcal{T}$ of subsets of X consisting of $\emptyset$, all open intervals (a, b), and all unions of open intervals is a topology on $\mathbb{R}$. It is called the **standard** or **usual topology** on $\mathbb{R}$. $\square$

Example 5.1.4. Let $X = \mathbb{R}$ again, but this time consider the collection $\mathcal{T}$ of subsets of X consisting of $\emptyset$, all intervals of the form $[a, b)$, and all unions of intervals of the form $[a, b)$. This is also a topology on $\mathbb{R}$. It is called the **lower-limit topology** on $\mathbb{R}$. Notice that the interval $[2, 5)$ is an open set in the lower-limit topology on $\mathbb{R}$, but $[2, 5)$ is not an open set in the usual topology on $\mathbb{R}$. However, the interval $(2, 5)$ is an open set in the usual topology *and* in the lower-limit topology. To see that $(2, 5)$ is an open set in the lower-limit topology, it suffices to note how we can express $(2, 5)$ as the union of intervals of the form $[a, b)$:

$$(2, 5) = \bigcup_{n=1}^{\infty} [2 + \frac{1}{n}, 5).$$

$\square$

Example 5.1.5. Let X be the set $\{a, b, c\}$. Let $\mathcal{T}$ be the set $\{\emptyset, \{a\}, \{a, b\}, X\}$, and let $\mathcal{S}$ be the set $\{\emptyset, \{a\}, \{b\}, X\}$. Then $\mathcal{T}$ is a topology on X, since it satisfies the three conditions of Definition 5.1.1. However, $\mathcal{S}$ is not a topology on X, since it fails the third condition: the union of the two elements $\{a\}$ and $\{b\}$ of $\mathcal{S}$ is not an element of $\mathcal{S}$ itself. $\square$

Example 5.1.6. Let $(X, \mathcal{T})$ be a topological space and let $Y \subseteq X$. Then the collection

$$\mathcal{T}_Y = \{A \cap Y : A \in \mathcal{T}\}$$

is a topology on Y, called the **relative** or **subspace** topology on Y.

For a specific example, let $X = \mathbb{R}$ and let $\mathcal{T}$ be the usual topology on $\mathbb{R}$. Let $Y = [0, 3] \cup \{5\}$. Then $\{5\}$ is not an open set in $\mathcal{T}$, but $\{5\}$ *is* an open set in $\mathcal{T}_Y$, since $\{5\} = (4, 6) \cap Y$ and $(4, 6) \in \mathcal{T}$. Also, $[0, 1)$ is open in $\mathcal{T}_Y$, but $[0, 1)$ is not open in $\mathcal{T}$.   □

Example 5.1.7. Let X be the plane $\mathbb{R}^2$. Then the ϵ-ball around a point (x, y), denoted by $B_\epsilon((x, y))$, is the set of all points in the plane whose distance to (x, y) is less than ϵ:

$$B_\epsilon((x, y)) = \{(a, b) \in \mathbb{R}^2 : \sqrt{(a - x)^2 + (b - y)^2} < \epsilon\}.$$

The collection of subsets of X consisting of the empty set, all ϵ-balls around all points in the plane, and all unions of ϵ-balls in the plane is a topology on $\mathbb{R}$, called the **usual topology** on $\mathbb{R}^2$.

The circle is a subset of the plane $\mathbb{R}^2$, and as such we can form the subspace topology on the circle. This subspace topology is referred to as the usual topology on the circle. Similarly, we can define the usual topology on the square or on the triangle, or on any figure in the plane.   □

Definition 5.1.8. Let $(X, \mathcal{T})$ and $(Y, \mathcal{S})$ be topological spaces. Then a function $f : X \to Y$ is **continuous** if for every set U which is open in $\mathcal{S}$, the inverse image $f^{-1}(U)$ is an open set in $\mathcal{T}$.

In calculus you may have seen a different definition of continuous: the function $f : \mathbb{R} \to \mathbb{R}$ is *continuous at* $x = a$ if

$$\lim_{x \to a} f(x) = f(a);$$

and f is continuous on $\mathbb{R}$ if it is continuous at every $a \in \mathbb{R}$. The topological definition of continuous in Definition 5.1.8 is a generalization of this definition; that is, a function $f : \mathbb{R} \to \mathbb{R}$ is continuous using the calculus definition if and only if the same function $f : \mathbb{R} \to \mathbb{R}$ is continuous using Definition 5.1.8, where the topology on $\mathbb{R}$ is the usual topology. Indeed, the subject of topology partly originated as an attempt to generalize the notion of continuity to functions between sets other than $\mathbb{R}$. Since continuity plays such a big part in calculus, one might expect continuity to play a big part in a more general setting; indeed, it does.

Proposition 5.1.9. *Let $(X, \mathcal{T})$, $(Y, \mathcal{S})$ and $(Z, \mathcal{V})$ be topological spaces. Suppose $f : X \to Y$ and $g : Y \to Z$ are continuous. Then the composition $g \circ f : X \to Z$ is continuous.*

The proof of Proposition 5.1.9 is Exercise 5.1.12.

Definition 5.1.10. Let $(X, \mathcal{T})$ and $(Y, \mathcal{S})$ be topological spaces. Then a function $f : X \to Y$ is a **homeomorphism** if f is one-to-one, onto, continuous, and if f^{-1} is continuous. The spaces $(X, \mathcal{T})$ and $(Y, \mathcal{S})$ are called **homeomorphic** if there is a homeomorphism $f : X \to Y$.

When we say that two topological spaces are topologically the same, it means that they are homeomorphic.

Example 5.1.11. Let X be the set $\{a, b\}$, and let $\mathcal{T}$ be the topology on X given by $\mathcal{T} = \{\emptyset, \{a\}, X\}$. Let $Y = \{1, 2\}$, and let $\mathcal{S}$ be the topology on Y given by $\{\emptyset, \{1\}, Y\}$. Then the topological spaces $(X, \mathcal{T})$ and $(Y, \mathcal{S})$ are homeomorphic.

To prove that $(X, \mathcal{T})$ and $(Y, \mathcal{S})$ are homeomorphic, it suffices to find a function $f : X \to Y$ which is a homeomorphism. Define $f : X \to Y$ by $f(a) = 1$, $f(b) = 2$. It is easy to check that f is one-to-one and onto. To see that f is continuous, note that the only open sets in Y are $\emptyset$, $\{1\}$, and Y. Since $f^{-1}(\emptyset) = \emptyset$, $f^{-1}(\{1\}) = \{a\}$, and $f^{-1}(Y) = X$, and $\emptyset$, $\{a\}$, and X are all open sets in X, the inverse image of every open set in Y is an open set in X. Thus, f is continuous. Similarly, f^{-1} is continuous since the image of every open set in X is an open set in Y. Since f is one-to-one, onto, continuous, and has a continuous inverse, f is a homeomorphism.

□

Example 5.1.12. Let S^1 denote the unit circle $\{(x, y) : x^2 + y^2 = 1\}$. Let Q denote the square

$\{(x, y) : \text{ either } x = \pm 1 \text{ and } -1 \le y \le 1, \text{ or } y = \pm 1 \text{ and } -1 \le x \le 1\}$.

Consider S^1 and Q as subspaces of the usual topology on the plane. Then the function $f : Q \to S^1$ given by

$$f((x, y)) = \left(\frac{x}{\sqrt{x^2 + y^2}}, \frac{y}{\sqrt{x^2 + y^2}} \right)$$

is a homeomorphism. The function f takes a point on the square to the point on the circle where the line segment from the origin to the point on the square intersects the circle; see Figure 5.1.

□

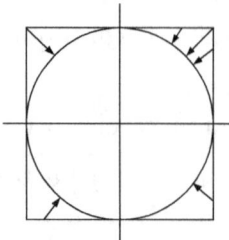

Figure 5.1: A homeomorphism from a square to a circle

Exercises

5.1.1. Show that the lower-limit topology on $\mathbb{R}$ really is a topology on $\mathbb{R}$.

5.1.2. Let $X = \{a, b, c\}$. Which of the following are topologies on X?
(a) $\{\emptyset, \{a\}, X\}$
(b) $\{\emptyset, \{a\}, \{b\}, \{c\}, X\}$
(c) $\{\emptyset, \{a\}, \{a, b\}, X\}$
(d) $\{\emptyset, \{a, b\}, X\}$
(e) $\{\emptyset, \{a, b\}, \{b, c\}, X\}$
(f) $\{\emptyset, \{a\}, \{b, c\}, X\}$
(g) $\{\emptyset, \{a\}, \{b\}, \{a, b\}, \{b, c\}, X\}$

5.1.3. Let $\mathcal{T}$ and $\mathcal{S}$ be topologies on a set X.
(a) Prove that $\mathcal{T} \cap \mathcal{S}$ is a topology on X.
(b) Prove that $\mathcal{T} \cup \mathcal{S}$ is not necessarily a topology on X.

5.1.4. Let $\mathcal{T} = \{\emptyset, \mathbb{R}\} \cup \{(a, b) : a < b\}$. Prove that $\mathcal{T}$ is not a topology on $\mathbb{R}$.

5.1.5. Show that the set $\{(x, y) : -1 < x < 1, \ -1 < y < 1\}$ is an open set in the usual topology on $\mathbb{R}^2$.

5.1.6. Show that the set $\{(x, y) : -1 \le x \le 1, \ -1 \le y \le 1\}$ is not an open set in the usual topology on $\mathbb{R}^2$.

5.1.7. (a) Show that the set $[1, 4]$ is a closed set in the usual topology on $\mathbb{R}$.
(b) Show that $[1, 4]$ is a closed set in the discrete topology on $\mathbb{R}$.
(c) Show that $[1, 4]$ is closed in the lower-limit topology on $\mathbb{R}$.
(d) Show that $[1, 4]$ is not closed in the trivial topology on $\mathbb{R}$.

5.1.8. Let $X = \{a, b, c\}$ and $\mathcal{T} = \{\emptyset, \{a\}, \{a, b\}, X\}$. (Then $\mathcal{T}$ is a topology on X.) List all of the closed sets in X.

5.1.9. Let $Y = [1, 4] \cup [11, 14]$. Let $\mathcal{T}$ be the usual topology on $\mathbb{R}$, and let $\mathcal{T}_Y$ be the subspace topology on Y. Show that $[1, 4]$ and $[11, 14]$ are both open and closed in this topology.

5.1.10. Let X be a set with the discrete topology, and let $(Y, \mathcal{S})$ be a topological space. Prove that every function $f : X \to Y$ is continuous.

5.1.11. Let Y be a set with the trivial topology, and let $(X, \mathcal{T})$ be a topological space. Prove that every function $f : X \to Y$ is continuous.

5.1.12. Prove Proposition 5.1.9.

5.1.13. Let A be a set of topological spaces. Define a relation R on A by $(X, \mathcal{T}) \, R \, (Y, \mathcal{S})$ if there is a homeomorphism $f : X \to Y$. Prove that R is an equivalence relation on A.

5.1.14. Give the intervals $(1, 2)$ and $(1, 5)$ the subspace topologies of the usual topology on $\mathbb{R}$. Find a homeomorphism $f : (1, 2) \to (1, 5)$.

5.1.15. Give the intervals (a, b) and (c, d) the subspace topologies of the usual topology on $\mathbb{R}$. Find a homeomorphism $f : (a, b) \to (c, d)$.

5.1.16. Give $(0, 1)$ and $(-\pi/2, \pi/2)$ the subspace topologies of the usual topology on $\mathbb{R}$, and let $\mathbb{R}$ have the usual topology.
(a) Find a homeomorphism $f : (-\pi/2, \pi/2) \to \mathbb{R}$.
(b) Find a homeomorphism $f : (0, 1) \to \mathbb{R}$.

5.1.17. The **upper-limit topology** on $\mathbb{R}$ consists of the empty set, $\mathbb{R}$, and all unions of intervals of the form $(a, b]$. Find a homeomorphism from the lower-limit topology to the upper-limit topology.

5.2 Topological properties

In order to prove that two topological spaces are homeomorphic, it is enough to find a function which turns out to be a homeomorphism. However, to show that two topological spaces are not homeomorphic, it is not enough to define a function which turns out not to be a homeomorphism. Indeed, some other function might be a homeomorphism. To show that two spaces are not homeomorphic, it is enough to prove that no function between the spaces can possibly be a homeomorphism. This can be quite difficult at times. Instead, we usually look for a property that one space has that the other does not; but we only look for those properties that are invariant under homeomorphisms.

Definition 5.2.1. A **topological property** of a topological space $(X, \mathcal{T})$ is a characteristic of the space $(X, \mathcal{T})$ such that whenever another topological space $(Y, \mathcal{S})$ is homeomorphic to $(X, \mathcal{T})$, $(Y, \mathcal{S})$ must have the same characteristic.

Theorem 5.2.2. *Let $(X, \mathcal{T})$ be a topological space. Then the cardinality $|X|$ is a topological property. That is, if $(X, \mathcal{T})$ and $(Y, \mathcal{S})$ are homeomorphic, then $|X| = |Y|$.*

Proof. Suppose that $(X, \mathcal{T})$ and $(Y, \mathcal{S})$ are homeomorphic. Then there is a homeomorphism $f : X \to Y$. Since f is a homeomorphism, f is a bijection. By Definition 4.5.1, $|X| = |Y|$.   **q.e.d.**

Another topological property is the notion of *connectedness*. Imprecisely, a topological space is connected if it is not naturally composed of more than one part.

Definition 5.2.3. Let $(X, \mathcal{T})$ be a topological space. Then X is **connected** if there do not exist two nonempty, disjoint, open subsets U, V of X such that $U \cup V = X$. The space X is **disconnected** if it is not connected.

Example 5.2.4. Consider the set $X = (0, 1) \cup (3, 4)$ with the subspace topology of the usual topology on $\mathbb{R}$. Then the subsets $U = (0, 1)$ and $V = (3, 4)$ are nonempty, disjoint, open subsets of X whose union is all of X. Hence, X is not connected.   □

Theorem 5.2.5. *Connectedness is a topological property.*

The proof of Theorem 5.2.5 is Exercise 5.2.8.

Example 5.2.6. The topological space $\mathbb{R}$ with the lower-limit topology is not homeomorphic to $\mathbb{R}$ with the usual topology. In the space $\mathbb{R}$ with the lower-limit topology, the sets $U = (-\infty, 0)$ and $V = [0, \infty)$ are nonempty, disjoint, open subsets of $\mathbb{R}$ whose union is all of $\mathbb{R}$. Hence, $\mathbb{R}$ with the lower-limit topology is not connected. However, the space $\mathbb{R}$ with the usual topology is connected because there are no nonempty, disjoint, open subsets of $\mathbb{R}$ in the usual topology whose union is all of $\mathbb{R}$. (Proving that there are no such subsets is not easy, but intuitively, open sets are unions of open intervals, and endpoints of these intervals would have to be contained in one of the subsets.) Since one space is connected and the other is not, the two spaces are not homeomorphic. □

Definition 5.2.7. Let $(X, \mathcal{T})$ be a connected topological space. Then the point $a \in X$ is a **cut point** of X if the subspace $X - \{a\}$ is disconnected.

Example 5.2.8. The point 0 is a cut point of $\mathbb{R}$ with the usual topology. Since $\mathbb{R} - \{0\} = (-\infty, 0) \cup (0, \infty)$, the sets $U = (-\infty, 0)$ and $V = (0, \infty)$ form nonempty, disjoint, open sets whose union is all of $\mathbb{R} - \{0\}$. In fact, every point of $\mathbb{R}$ is a cut point of $\mathbb{R}$ with the usual topology. □

Theorem 5.2.9. *Let $(X, \mathcal{T})$ and $(Y, \mathcal{S})$ be topological spaces. Suppose $f : X \to Y$ is a homeomorphism. Then if a is a cut point of X, then $f(a)$ is a cut point of Y.*

Proof. Suppose $f : (X, \mathcal{T}) \to (Y, \mathcal{S})$ is a homeomorphism and a is a cut point of X. We want to show that $Y - \{f(a)\}$ is disconnected. Since a is a cut point of X, the subspace $X - \{a\}$ is disconnected. Hence, there are two nonempty, disjoint, open sets U and V in the subspace $X - \{a\}$ such that $U \cup V = X - \{a\}$. Since f is a homeomorphism, it is easy to see that (1) $f(U)$ and $f(V)$ are nonempty; (2) $f(U)$ and $f(V)$ are disjoint; (3) $f(U)$ and $f(V)$ are open in $Y - \{f(a)\}$; and (4) $f(U) \cup f(V) = Y - \{f(a)\}$. Thus, $f(a)$ is a cut point of $Y - \{f(a)\}$. **q.e.d.**

Example 5.2.10. The circle $S^1 = \{(x, y) : x^2 + y^2 = 1\}$, with the subspace topology as a subspace of the plane with the usual topology, is not homeomorphic to $\mathbb{R}$ with the usual topology. To see this, suppose S^1 and $\mathbb{R}$ are homeomorphic. Then there exists a homeomorphism $f : \mathbb{R} \to S^1$. Since the point 0 is a cut point of $\mathbb{R}$, the point $f(0)$ must be a cut point of S^1. However, S^1 has no cut points because deleting a point from the circle S^1 yields a connected space. Therefore, by contradiction, S^1 and $\mathbb{R}$ cannot be homeomorphic. □

We began our study of topology in Section 5.1 by making the statement that a circle can be stretched and bent to form an ellipse, triangle or square, but that it cannot be stretched into the shape of a line or a pair of disjoint circles. This can now be demonstrated using homeomorphisms (as in Example 5.1.12), cut points (as in Example 5.2.10) and connectedness (as in Exercise 5.2.9).

Exercises

5.2.1. Let $X = \{a, b, c, d, e\}$ and $Y = \{\alpha, \beta, \gamma, \delta\}$. Let $\mathcal{T} = \{\emptyset, \{a\}, X\}$ and $\mathcal{S} = \{\emptyset, \{\alpha\}, Y\}$. Prove that $(X, \mathcal{T})$ and $(Y, \mathcal{S})$ are not homeomorphic.

5.2.2. Let $X = \{1, 2, 3\}$, $Y = \{a, b, c\}$, $\mathcal{T} = \{\emptyset, \{a\}, \{b, c\}, X\}$, and $\mathcal{S} = \{\emptyset, \{1\}, \{1, 2\}, Y\}$. Prove that $(X, \mathcal{T})$ and $(Y, \mathcal{S})$ are not homeomorphic.

5.2.3. Let $X = \{1, 2, 3, 4\}$ and let $\mathcal{T} = \{\emptyset, X, \{1\}, \{4\}, \{1, 4\}, \{2, 3, 4\}\}$. Then $\mathcal{T}$ is a topology on X. Show that $(X, \mathcal{T})$ is disconnected.

5.2.4. Let $X = \{1, 2, 3\}$ and let $\mathcal{T} = \{\emptyset, X, \{1\}, \{1, 2\}\}$. Then $\mathcal{T}$ is a topology on X. Show that $(X, \mathcal{T})$ is connected.

5.2.5. Prove that a topological space $(X, \mathcal{T})$ is disconnected if and only if there is a nonempty, proper subset of X which is both open and closed.

5.2.6. Prove that the subspace $X = \{1, 2\}$ of $\mathbb{R}$ with the usual topology is not connected.

5.2.7. Prove that the subspace $\mathbb{Q}$ of $\mathbb{R}$ with the usual topology is not connected.

5.2.8. Prove Theorem 5.2.5.

5.2.9. Prove that, as subspaces of the plane with the usual topology, a circle is not homeomorphic to a pair of disjoint circles.

5.2.10. Prove that, with the usual topologies, $\mathbb{R}$ and $\mathbb{R}^2$ are not homeomorphic.

5.2.11. Let X be any set, and give X the trivial topology. Show that X has no cut points.

5.2.12. Let $X = \{\alpha, \beta, \gamma\}$, and let $\mathcal{T}$ be the topology $\{\emptyset, \{\alpha\}, \{\beta\}, \{\alpha, \beta\}, X\}$. Determine which of the points α, β, γ is a cut point of X.

5.3 Abstract algebra, groups and isomorphisms

Mathematics is about abstraction and generalization. **Abstraction** is the process of formulating general concepts by studying specific examples and extracting features that these examples have in common. The abstraction of a concept might detach it from any specific example from which it came and generalize the concept so that the applications are more extensive.

As an example, consider the problem of determining what numerical grade you might need on a final exam in order to make an A in your biology class. After solving this type of problem for each of your classes you will find that the solution amounts to solving a linear equation. The general concept of solving linear equations is detached from the original concept of figuring out a grade necessary on a final exam in biology, but the general concept can be used in many more applications.

Abstract algebra is an abstraction of the concept of arithmetic. The operations of addition and multiplication are such essential accessories to the sets of integers or real numbers that these operations give these sets a mathematical form called an **algebraic structure**. Other algebraic structures can be found in other sets with other operations, and the subject of abstract algebra is the study of these algebraic structures.

Definition 5.3.1. A **binary operation** on a set S is a function whose domain is $S \times S$ and whose codomain is S.

Example 5.3.2. The function $f : \mathbb{Z} \times \mathbb{Z} \to \mathbb{Z}$ given by $f(n,m) = n + m$ is a binary operation on $\mathbb{Z}$. That is, addition is a binary operation on the set of integers. Rather than writing $f(n,m)$, we usually just write $n + m$.

In the same way, multiplication is a binary operation on $\mathbb{Z}$. Subtraction is a binary operation on $\mathbb{Z}$, but subtraction is not a binary operation on $\mathbb{N}$, since it is possible to subtract one natural number from another and not end up with a natural number. Division is not a binary operation on $\mathbb{R}$ because division by zero is undefined. Division is a binary operation on $\mathbb{R}^+$. $\square$

Example 5.3.3. Let $\mathcal{F}$ denote the set of functions $f : \mathbb{R} \to \mathbb{R}$. Then composition of functions is a binary operation on $\mathcal{F}$. Similarly, composition is a binary operation on the set of continuous functions from $\mathbb{R}$ to $\mathbb{R}$, and it is a binary operation on the set of differentiable functions from $\mathbb{R}$ to $\mathbb{R}$. $\square$

A binary operation $f : A \times A \to A$ is usually denoted by a single symbol like $*$, so that $a * b$ denotes $f(a,b)$. The binary operation of addition on $\mathbb{Z}$

is denoted by $n + m$; multiplication on $\mathbb{R}$ is denoted by $a \cdot b$; division on $\mathbb{R}^+$ is denoted by a/b or $a \div b$; and composition on a set of functions is denoted by $g_1 \circ g_2$.

It is not simply the operations alone on a set that give the set an algebraic structure, but it is also the properties of the operations that hold on the set. For example, addition on the set of integers is associative, meaning that $(n + m) + k = n + (m + k)$ for all integers n, m, and k.

Definition 5.3.4. A **group** is a set G along with a binary operation $*$ on G such that the following properties hold.

1. (Associativity) For all $a, b, c \in G$, $(a * b) * c = a * (b * c)$.

2. (Existence of an identity) There is an element $e \in G$, called the **iden-tity** element of G, such that for all $a \in G$, $e * a = a * e = a$.

3. (Existence of inverses) For each $a \in G$ there is a corresponding $b \in G$ (called the **inverse** of a) such that $a * b = b * a = e$.

Since a group consists of a set G along with an operation $*$, the group is denoted by the pair $(G, *)$.

Example 5.3.5. $(\mathbb{Z}, +)$ is a group. In other words, the integers form a group under addition. To check this, we would need to check first that $+$ is a binary operation on $\mathbb{Z}$. Since the sum of any two integers is an integer, addition is indeed a binary operation on $\mathbb{Z}$. Next, we would check that addition is associative on $\mathbb{Z}$. The identity element e is the integer 0, since $0 + n = n + 0 = n$ for every integer n. Finally, the inverse of any integer n is the integer $-n$. $\qquad \square$

Example 5.3.6. The set of positive rational numbers is a group under mul-tiplication. This group is denoted by $(\mathbb{Q}^+, \cdot)$. To see that this is a group, we first note that the product of two positive rational numbers is another positive rational number. Thus, multiplication is a binary operation on $\mathbb{Q}^+$. Clearly associativity holds for multiplication of positive rationals. The iden-tity in this group is $e = 1$, since $1 \cdot \frac{p}{q} = \frac{p}{q} \cdot 1 = \frac{p}{q}$ for every $\frac{p}{q} \in \mathbb{Q}^+$. Finally, the inverse of the positive rational number $\frac{p}{q}$ is the positive rational number $\frac{q}{p}$. (Note that if $\frac{p}{q} \in \mathbb{Q}^+$, then $p \neq 0$ and $q \neq 0$.) $\qquad \square$

Example 5.3.7. Let T denote the set $\{2^n : n \in \mathbb{Z}\}$. Define a binary operation $*$ on T by $2^k * 2^\ell = 2^{k+\ell}$. Then $(T, *)$ is a group. Associativity is easy to verify, as $(2^k * 2^\ell) * 2^m = 2^{k+\ell} * 2^m = 2^{(k+\ell)+m} = 2^{k+(\ell+m)} = 2^k * 2^{\ell+m} = 2^k * (2^\ell * 2^m)$. The identity in this group is the element $2^0 \in T$, since $2^0 * 2^k = 2^k * 2^0 = 2^k$ for all $2^k \in T$. The inverse of any element 2^k is 2^{-k}. □

Example 5.3.8. Let $S = \mathcal{P}(\{1, 2, 3, ..., n\})$ denote the power set of the set of positive integers from 1 to n. Is S a group under the union ($\cup$) operation?

Solution. We go through the list of properties of a group, checking that our operation is a binary operation, that the operation is associative, that an identity element exists, and that inverses of all elements exist. First, let $A, B \in S$. Then A and B are subsets of $\{1, 2, 3, ..., n\}$, and so $A \cup B$ is also a subset of $\{1, 2, 3, ..., n\}$. Hence, $A \cup B \in S$, and this is a binary operation on S.

Second, we saw in part (2)(a) of Theorem 3.2.2 that the associative property holds for unions of sets.

Third, the empty set works as the identity element in S since $\emptyset \cup A = A \cup \emptyset = A$ for every $A \in S$.

Fourth, we check for the existence of inverses. Given any $A \in S$, we would like to find an element $B \in S$ such that $A \cup B = \emptyset$, since the identity is $\emptyset$. Unfortunately, most elements of S do not have inverses, since $A \cup B$ must contain A, which in general will not be empty.

Since not every element of S has an inverse, the set S is not a group under the union operation. □

Example 5.3.9. Let n be a positive integer, and let $\mathbb{Z}_n$ denote the set of equivalence classes of the modular congruence equivalence relation $\equiv$ (mod n). Then $\mathbb{Z}_n = \{[0], [1], \ldots, [n-1]\}$. Define a binary operation $+_n$ on $\mathbb{Z}_n$ by $[x] +_n [y] = [x+y]$. It follows from Exercise 3.4.10 that this operation is *well-defined*. What **well-defined** means in this context is that if we choose different representatives of the same equivalence classes, the result under the operation remains the same. In other words, this means that if $[x] = [z]$ and $[y] = [w]$, then $[x+y] = [z+w]$.

The set $\mathbb{Z}_n$ is a group under the operation $+_n$; the proof is Exercise 5.3.6. □

Example 5.3.10. For $n \in \mathbb{N}$, let $S = \{1, 2, 3, \ldots, n\}$, and let S_n denote the set of bijections from S to S. (Each such bijection $f : S \to S$ is called a **permutation** of S.) Then composition $\circ$ is a binary operation on S_n because the composition of two bijections is a bijection. (See Theorems 4.1.20 and 4.1.21.) Furthermore, $(S_n, \circ)$ is a group. To prove this, first note that associativity holds for compositions of all functions, and so associativity must hold for compositions of bijections from S to S. Next, note that the identity function $\mathrm{id}_S : S \to S$ given by the formula $\mathrm{id}_S(j) = j$ for all $j \in S$ acts as the identity e in this group, since $\mathrm{id}_S \circ f = f \circ \mathrm{id}_S = f$ for all $f \in S_n$. Finally, by Theorem 4.2.12, each bijection $f : S \to S$ has an inverse function $f^{-1} : S \to S$ which is also a bijection, and $f \circ f^{-1} = f^{-1} \circ f = \mathrm{id}_S$.

The group $(S_n, \circ)$ is called the **symmetric group on** S. □

Definition 5.3.11. Let $(G, *_1)$ and $(H, *_2)$ be groups. Then a function $f : G \to H$ is an **isomorphism** if f is one-to-one, onto, and

$$f(a *_1 b) = f(a) *_2 f(b)$$

for all $a, b \in G$. The groups $(G, *_1)$ and $(H, *_2)$ are said to be **isomorphic** if there is an isomorphism $f : G \to H$. Any function $f : G \to H$ satisfying

$$f(a *_1 b) = f(a) *_2 f(b)$$

for all $a, b \in G$ is said to be **operation-preserving**. Thus, an isomorphism is an operation-preserving bijection.

When we say that two groups are algebraically the same, we mean that they are isomorphic.

Example 5.3.12. Consider the two groups $(\mathbb{Z}, +)$ and $(T, *)$ given in Examples 5.3.5 and 5.3.7. Define a function $f : \mathbb{Z} \to T$ by $f(m) = 2^m$. Then f is an isomorphism. It is easy to verify that f is a bijection. To verify that f is operation-preserving, we note that $f(m+n) = 2^{m+n} = 2^m * 2^n = f(m) * f(n)$ for all $m, n \in \mathbb{Z}$. □

Exercises

5.3.1. Is the set of even integers a group under addition? Prove your answer.

5.3.2. Is the set of integers a group under subtraction? Prove your answer.

5.3.3. Is the set of integers a group under multiplication? Prove your answer.

5.3.4. Let $G = \{-1, 1\}$. Let $\cdot$ denote the operation of multiplication on G. Prove that $(G, \cdot)$ is a group.

5.3.5. Does the formula $[x] * [y] = [x^2 + y^2]$ determine a well-defined binary operation $*$ on $\mathbb{Z}_n$? Prove your answer.

5.3.6. Prove that $(\mathbb{Z}_n, +_n)$ is a group.

5.3.7. List all the elements of the symmetric group S_3.

5.3.8. Let n be an integer greater than 1. Define a binary operation $\odot$ on $\mathbb{Z}_n$ by $[x] \odot [y] = [xy]$. Prove that:
(a) the binary operation $\odot$ is well-defined; but
(b) $(\mathbb{Z}_n, \odot)$ is not a group.

5.3.9 (Linear Algebra required). Prove that the set of all invertible 2×2 matrices with entries in $\mathbb{R}$ forms a group under matrix multiplication.

5.3.10. Let $2\mathbb{Z}$ denote the set of even integers. Then $(2\mathbb{Z}, +)$ is a group, as you may have discovered in Exercise 5.3.1. Prove that $(2\mathbb{Z}, +)$ is isomorphic to $(\mathbb{Z}, +)$.

5.3.11. Let A be a set of groups. Define a relation R on A by $(G, *_1) \, R \, (H, *_2)$ if there is an isomorphism $f : G \to H$. Prove that R is an equivalence relation on A.

5.3.12. What are some similarities and differences between isomorphisms and homeomorphisms?

5.4 Algebraic properties

In order to prove that two groups are isomorphic, it suffices to find a function from one group to the other that turns out to be an isomorphism. But to prove that two groups are not isomorphic, we must prove that no such function exists. Often we can achieve this by finding a property that one group has but the other group does not; but we must make sure that the property we find is invariant under isomorphism. That is, if the group $(G, *_1)$ has the property and $(H, *_2)$ is isomorphic to $(G, *_1)$, then $(H, *_2)$ must have the property also. We start with cardinality.

Theorem 5.4.1. *Suppose two groups* $(G, *_1)$ *and* $(H, *_2)$ *are isomorphic. Then* $|G| = |H|$.

Proof. Suppose the groups $(G, *_1)$ and $(H, *_2)$ are isomorphic. Then there exists an isomorphism $f : G \to H$. By the definition of isomorphism, the function f must be a bijection. Hence, $|G| = |H|$.   **q.e.d.**

Definition 5.4.2. The **order** of a group G is the cardinality $|G|$. The group G is called **finite** if $|G|$ is finite; otherwise, G is called **infinite**.

Example 5.4.3. The order of $(\mathbb{Z}_n, +_n)$ is n. Hence, if $n \neq m$ then $(\mathbb{Z}_n, +_n)$ cannot be isomorphic to $(\mathbb{Z}_m, +_m)$. Moreover, since $(\mathbb{Z}, +)$ is an infinite group and for each n $(\mathbb{Z}_n, +_n)$ is a finite group, $(\mathbb{Z}, +)$ cannot be isomorphic to $(\mathbb{Z}_n, +_n)$ for any n.   □

Definition 5.4.4. Let $(G, *)$ be a group. Then $(G, *)$ is called **Abelian** (after the mathematician Niels Henrik Abel, 1802-1829) if G is commutative under the operation $*$; that is, if for all $a, b \in G$, $a * b = b * a$.

Example 5.4.5. The group $(\mathbb{Z}, +)$ is Abelian because $n + m = m + n$ for every pair of integers n and m. The group $(T, *)$ defined in Example 5.3.7 is also Abelian since $2^k * 2^\ell = 2^{k+\ell} = 2^{\ell+k} = 2^\ell * 2^k$ for all 2^k and 2^ℓ in T.   □

Example 5.4.6. For each $n \in \mathbb{N}$, the group $(\mathbb{Z}_n, +_n)$ is Abelian. To see this, note that for all $[x]$, $[y] \in \mathbb{Z}_n$, $[x] +_n [y] = [x + y] = [y + x] = [y] +_n [x]$.   □

Example 5.4.7. If $n \geq 3$ then the symmetric group $(S_n, \circ)$ is not Abelian. To prove this, let $n \geq 3$, $S = \{1, 2, \ldots, n\}$, and let $f : S \to S$ be the function defined by the formula

$$f(j) = \begin{cases} 2 & \text{if } j = 1 \\ 3 & \text{if } j = 2 \\ 1 & \text{if } j = 3 \\ j & \text{if } j > 3. \end{cases}$$

Then f is a bijection and so f is an element of S_n. Likewise, let $g : S \to S$ be given by

$$g(j) = \begin{cases} 3 & \text{if } j = 1 \\ 2 & \text{if } j = 2 \\ 1 & \text{if } j = 3 \\ j & \text{if } j > 3. \end{cases}$$

Then g is also an element of S_n. Now the compositions $f \circ g$ and $g \circ f$ are also elements of S_n, but $f \circ g \neq g \circ f$ because $f \circ g(1) = f(g(1)) = f(3) = 1$ but $g \circ f(1) = g(f(1)) = g(2) = 2$. Hence, for $n \neq 3$, $(S_n, \circ)$ is not Abelian. $\square$

Theorem 5.4.8. *If the group $(G, *_1)$ is Abelian and the groups $(G, *_1)$ and $(H, *_2)$ are isomorphic, then $(H, *_2)$ is Abelian.*

Proof. Suppose $(G, *_1)$ is Abelian and $(G, *_1)$ and $(H, *_2)$ are isomorphic. Then there exists an isomorphism $f : (G, *_1) \to (H, *_2)$. Suppose $h_1, h_2 \in H$. Then since f is onto, there are elements $g_1, g_2 \in G$ such that $f(g_1) = h_1$ and $f(g_2) = h_2$. Hence,

$$\begin{aligned}
h_1 *_2 h_2 &= f(g_1) *_2 f(g_2) \\
&= f(g_1 *_1 g_2) \text{ (because f is orientation-preserving)} \\
&= f(g_2 *_1 g_1) \text{ (because $(G, *_1)$ is Abelian)} \\
&= f(g_2) *_2 f(g_1) \text{ (because f is orientation-preserving)} \\
&= h_2 *_2 h_1.
\end{aligned}$$

q.e.d.

Example 5.4.9. For any $n \in \mathbb{N}$ and any integer $m \geq 3$, the groups $(\mathbb{Z}_n, +_n)$ and $(S_m, \circ)$ cannot be isomorphic because for $m \geq 3$ none of the symmetric groups $(S_m, \circ)$ are Abelian but all of the groups $(\mathbb{Z}_n, +_n)$ are Abelian. $\square$

Exercises

5.4.1. Let $\mathbb{R}^+$ denote the set of positive real numbers, and let $\cdot$ denote multiplication.
(a) Prove that $(\mathbb{R}^+, \cdot)$ is a group.
(b) Prove that $(\mathbb{R}, +)$ is a group.
(c) Prove that $f : (\mathbb{R}, +) \to (\mathbb{R}^+, \cdot)$ given by $f(x) = e^x$ is an isomorphism.
(d) Prove that $(\mathbb{R}^+, \cdot)$ is not isomorphic to $(\mathbb{Z}, +)$.

5.4.2. Find the order of the symmetric group $(S_n, \circ)$.

5.4.3. Prove that the symmetric group $(S_2, \circ)$ is Abelian.

5.4.4 (Linear Algebra required). Exercise 5.3.9 was to prove that the set of all invertible 2×2 matrices with entries in $\mathbb{R}$ forms a group under matrix multiplication. Prove that this group is not Abelian.

5.4.5. Let $f : (\mathbb{Z}, +) \rightarrow (\mathbb{Z}_n, +_n)$ be given by $f(j) = [j]$. Prove that f is operation-preserving but that f is not an isomorphism.

5.4.6. Congratulations on making it to the end of this text! I hope that you will continue your studies in mathematics; this is only the beginning.

Appendix A

Hints on selected exercises

1.1.3. A paradox is a declarative statement that cannot be true and cannot be false. Why can't the given statement be true? Why can't it be false?

1.1.12. The question is asking whether $p \rightarrow q$ and $q \rightarrow p$ have the same meaning.

1.2.13. Don't be afraid to make several conclusions. See Example 1.2.7. If you find that one or more of your conclusions is not actually true for all positive integers n, then you should point this out.

1.2.14. Note that 4 is not the sum of two odd prime integers. (Why not?)

1.3.4. (a) $\forall n \in \mathbb{Z} \ (n > 0)$

1.3.10. (a) $\forall x \in \mathbb{R} \ (x \leq 3)$

2.1.3. First, what *kinds* of numbers are the results? (Irrationals?) Second, can you predict the outcome?

2.1.4. Your conjecture could turn out to be what is known as the **3n+1 Conjecture** or the **Hailstone Conjecture**. You can find more information on the conjecture via a search on the internet.

2.1.9. The expression $d - \ell + r$ is called the **Euler characteristic**. Much more information on this can be found on the internet.

2.2.9. In a direct proof, we would suppose that nm is even, so that for some integer k, $nm = 2k$. But from that supposition, when we solve for n (or m), we get $n = 2k/m$ (or $m = 2k/n$). It is not clear that n or m is even from this, so the direct proof doesn't seem to work well. When one proof method does not flow easily, you might want to try another method instead of trying to force the first method to work.

A proof by contrapositive works well here. Remember to use DeMorgan's Laws to change the negation of the disjunction to a conjunction of negations.

2.2.12 Before you begin a proof, you should figure out what a *perfect square* is.

2.2.14. Think about even and odd integers.

2.3.3. A straightforward way would be to consider four cases. One case could be when $x \geq 0$ and $y < 0$.

2.4.6. Use induction on n, not r. Start by supposing that r is given and $r \neq 1$. Then prove the statement for this fixed value of r.

2.4.16. A good notation will make the problem easier. Let $D^n(f(x))$ denote the nth derivative of $f(x)$. Then use the fact that $D^{n+1}(x^{n+1}) = D^n(D^1(x^{n+1})) = D^n((n+1)x^n) = (n+1)D^n(x^n)$.

2.4.19. Use induction. For the inductive step, use integration by parts.

2.4.20. First prove that the formula works for $n = 0$ and $n = 1$ by just plugging those values in. Then use the Strong Form of Induction to prove that the formula holds for every integer $n \geq 2$. For the inductive step, use

$$\left(\frac{1+\sqrt{5}}{2}\right)^{n+1} = \left(\frac{1+\sqrt{5}}{2}\right)^2 \left(\frac{1+\sqrt{5}}{2}\right)^{n-1}$$

and

$$\left(\frac{1-\sqrt{5}}{2}\right)^{n+1} = \left(\frac{1-\sqrt{5}}{2}\right)^2 \left(\frac{1-\sqrt{5}}{2}\right)^{n-1}$$

and work from both sides of the equation.

2.4.21. In the inductive step, consider the case when the integer $k + 1$ is even and the case when $k + 1$ is odd.

2.4.23. Suppose $p(n)$ is false for some $n \geq m$. Then consider the set of integers at least as big as m for which $p(n)$ is false.

2.4.25. The proposition is obviously false, but this does not answer the question. There is one particular sentence in the argument that is not valid.

3.1.7. In the inductive step, let $|S| = k + 1$ and let $x \in S$. Then there are two types of subsets of S: those that contain x and those that don't. Show that there are 2^k that contain x and 2^k that don't. Then there will be $2^k + 2^k = 2^{k+1}$ subsets of S.

3.1.10. Use Exercise 3.1.7.

3.3.14. Use induction on k. Exercise 3.3.13 can be used in the inductive step.

4.2.6. To prove that two functions are equal, you can consider a function as a set of ordered pairs. Then to show that they are equal, you can show that each is a subset of the other.

4.3.4. Find critical points for $f(x)$ using the derivative to get

$$f(C) = [(2/\sqrt{3})^3 - 4(2/\sqrt{3}), 0].$$

4.4.6. To show that the intersection is $\{0\}$, show that (1) the number 0 is in the intersection; (2) no negative numbers are in the intersection; and (3) no positive numbers are in the intersection.

4.4.7. To show that a set is empty, proceed by contradiction.

4.5.1. Use Theorems 4.5.5 and 4.1.20.

4.5.7. (a) Try repeated squaring.

5.1.1. Use Definition 5.1.1.

Appendix B

Table of the Greek alphabet

name	character	capital	name	character	capital
alpha	α	A	nu	ν	N
beta	β	B	xi	ξ	Ξ
gamma	γ	Γ	omicron	o	O
delta	δ	Δ	pi	π	Π
epsilon	ϵ	E	rho	ρ	P
zeta	ζ	Z	sigma	σ	Σ
eta	η	H	tau	τ	T
theta	θ	Θ	upsilon	υ	Υ
iota	ι	I	phi	ϕ	Φ
kappa	κ	K	chi	χ	X
lambda	λ	Λ	psi	ψ	Ψ
mu	μ	M	omega	ω	Ω

Appendix C

Table of symbols

symbol	meaning	page
$\land$	conjunction	3
$\lor$	disjunction	4
$\sim$	negation	4
$\rightarrow$	conditional	6
$\leftrightarrow$	biconditional	7
$\Leftrightarrow$	logical equivalence	12
$\Rightarrow$	implication	14
$\therefore$	therefore	16
$\exists$	there exists	22
$\mathbb{R}$	set of real numbers	23
$\mathbb{Z}$	set of integers	23
$\mathbb{Q}$	set of rationals	23
$\mathbb{N}$	set of natural numbers	23
$\mathbb{C}$	set of complex numbers	23
$\mathbb{R}^+$	set of positive reals	23
$\mathbb{Q}^+$	set of positive rationals	23
$\in$	is an element of	23, 63
$\exists!$	there exists a unique	24
$\forall$	for all	25

symbol	meaning	page
$\mid$	divides	40
$\lvert x \rvert$	absolute value	45
$\sum$	summation	54
$\prod$	product	54
$n!$	factorial	57
$\notin$	is not an element of	64
$\{x, y\}$	set containing x and y	64
$\lvert S \rvert$	cardinality of the set S	64
$\emptyset$	the empty set	64
U	universal set	65
$\subseteq$	is a subset of	65
$\supseteq$	is a superset of	65
$\subsetneq$	is a proper subset of	65
$\subset$	subset or proper subset	65
$\mathcal{P}(A)$	power set of A	66
$A \cup B$	union of A and B	66
$A \cap B$	intersection of A and B	66
$A - B$	set difference of A and B	66
$\overline{A}$	complement of A	66
$A \triangle B$	symmetric difference	78
(a, b)	ordered pair	78
$A \times B$	Cartesian product	78
$\mathbb{R}^2$	the real plane	79
$(a_1, a_2, \ldots, a_n)$	n-tuple	79
$\prod A_j$	Cartesian product	79
$a \mathrel{R} b$	a is related to b	81
$a \equiv b \pmod{n}$	a is congruent to b modulo n	82
$[a]$	equivalence class of a	86
A/R	A modulo R	87
$f : A \to B$	function from A to B	95
$g \circ f$	composition of g with f	101
p_i	coordinate projection	104

symbol	meaning	page
χ_A	characteristic function	105
$f\vert_A$	restriction of a function	105
id_A	identity function on A	106
f^{-1}	inverse of function f	106
$\bigcup A_\lambda$	union of a collection of sets	115
$\bigcap A_\lambda$	intersection of a collection of sets	115
$[a, b]$	closed interval	116
(a, b)	open interval	116
$[a, b),\ (a, b]$	half-open interval	116
$(a, \infty),\ (-\infty, b)$	open half-line	116
$[a, \infty),\ (-\infty, b]$	closed half-line	116
$\bigcup \Gamma$	union of a collection of sets	118
$\bigcap \Gamma$	intersection of a collection of sets	118
$\aleph_0$	cardinality of $\mathbb{N}$	122
$\aleph_1$	smallest cardinality $> \aleph_0$	125
$\mathcal{T}$	a topology	130
$(X, \mathcal{T})$	a topological space	130
$\mathcal{D}$	the discrete topology	131
$(G, *)$	a group	140
$\mathbb{Z}_n$	modular group	141
S_n	symmetry group	142

Index

www.ingramcontent.com/pod-product-compliance
Lightning Source LLC
Chambersburg PA
CBHW080830220526
45467CB00008B/2242